**This book is due for return not later than the last date
stamped above, unless recalled sooner.**

... JUL 1981

Monographs of the Physiological Society No. 33

Physiology of the gastro-intestinal lymphatic system

PHYSIOLOGY OF THE GASTRO-INTESTINAL LYMPHATIC SYSTEM

J. A. BARROWMAN

Faculty of Medicine, Memorial University of Newfoundland
St John's, Newfoundland

CAMBRIDGE UNIVERSITY PRESS
Cambridge
London · New York · Melbourne

Published by the Syndics of the Cambridge University Press
The Pitt Building, Trumpington Street, Cambridge CB2 1RP
Bentley House, 200 Euston Road, London NW1 2DB
32 East 57th Street, New York, NY 10022, USA
296 Beaconsfield Parade, Middle Park, Melbourne 3206, Australia

First published 1978

Printed in Great Britain
by Redwood Burn Ltd., Trowbridge & Esher

Library of Congress Cataloguing in Publication Data
Barrowman, J. A.
Physiology of the gastro-intestinal lymphatic system.
(Monographs of the Physiological Society; no. 33)
Includes bibliographical references and index.
1. Lymphatics. 2. Digestive organs. I. Title.
II. Series: Physiological Society. Monographs;
no. 33. [DNLM: 1. Lymphatic system – Physiology.
2. Gastrointestinal system – Physiology. W1 M0569QW
no. 33 / WI102 B278p]
QP115.B37   616.4'2   77–22823
ISBN 0 521 21710 5

FOR JUDITH

Monographs of the Physiological Society

Members of Editorial Board: *A. D. M. Greenfield (Chairman); R. A. Gregory, D. K. Hill, P. B. C. Matthews, T. A. Sears*

PUBLISHED BY CAMBRIDGE UNIVERSITY PRESS

28 *M. J. Purves* The Physiology of the Cerebral Circulation, 1972
29 *D. McK. Kerslake* The Stress of Hot Environments, 1972
30 *M. R. Bennett* Autonomic Neuromuscular Transmission, 1972
31 *A. G. Macdonald* Physiological Aspects of Deep Sea Biology, 1975
32 *M. Peaker and J. L. Linzell* Salt Glands in Birds and Reptiles, 1975
33 *J. A. Barrowman* Physiology of the Gastro-intestinal Lymphatic System, 1978

PUBLISHED BY ACADEMIC PRESS

34 *C. G. Phillips and R. Porter* Corticospinal Neurones, Their Role in Movements, 1977

CONTENTS

Contents

*Plates 1.1–2.1 are between pp. 4 and 5, the
rest are between pp. 164 and 165.*

PREFACE

The formation and circulation of lymph has attracted less attention than the circulation of the blood. For many, lymph is a theoretical concept rather than a tangible reality. This must be largely due to the difficulties inherent in studying a rather inaccessible system of tenuous, delicate vessels. However, the essential role of the lymph circulation in the homeostasis of interstitial fluid, and ultimately of the tissue cells themselves, has been amply demonstrated. In this capacity, it forms a vital link in the extravascular circulation of plasma protein and fluid. In contrast to the blood circulation where pressures and flows are constantly regulated within fairly narrow limits, the lymphatics in their capacity as an overflow system transport variable amounts of lymph; these amounts depend on the rate of formation of tissue fluid in the area drained by a particular lymph vessel and reflect changes in regional blood flow or capillary permeability.

In addition to these general functions of the lymphatic system, there are certain roles peculiar to the lymphatics of various regions or organs. Examples of these include the highly specialised function of the intestinal lymphatic system in the transport of absorbed lipids, and the transport of macromolecules, such as protein or peptide hormones, from endocrine tissues to the blood. In such roles the lymph vessels of an organ or system take part in the specialised function of the tissue. The lymphatics of the alimentary tract are involved in several ways in the normal physiological processes of that system. The purpose of this book is to examine these roles as they are played in the normal activity of the gastro-intestinal tract, that is: digestion, absorption and, to some extent, metabolism of nutrients. Studies of pathological processes causing disorganisation of the gastro-intestinal lymphatics will also be considered where they are able to throw further light on the normal function of the system. The important role of lymphoid tissue in the body's defence mechanisms is not discussed except where this is related to digestive physiology.

I am grateful to my colleagues who have read and criticised the manuscript, in particular Dr Kenneth Roberts, Memorial University of Newfoundland. I am also indebted to many authors who have allowed me to reproduce material and to Marilyn Fost who has typed the manuscript.

J. A. BARROWMAN

St John's
Newfoundland

1 EARLY INVESTIGATIONS OF THE LYMPHATIC SYSTEM

Records show that there has been interest in the anatomy and physiology of the lymphatic system since antiquity. As in many branches of medical science, early investigation in these fields was exceedingly slow, but periodically the science of *lymphology* took sudden remarkable steps forward. Such steps occurred, for example, during the sixteenth and seventeenth centuries in European medical schools, notably those in Italy and Scandinavia. The inconspicuous and fragile nature of lymphatic vessels has been responsible for the slow progress in investigation of the function of the system and the important physiological roles of the lymphatic circulation have therefore been appreciated more recently than those of other biological systems.

While the phenomena of lymphoedema and elephantiasis were mentioned in very early Hebrew and Greek literature, the first direct references to the lymphatics were made by the Greeks in the fourth century B.C. At this time Aristotle may have observed lymphatics, since he described vessels containing a colourless fluid, while Hippocrates referred to 'white blood'. Clearer descriptions of lymph vessels, however, came from the Alexandrian school of medicine: in the fourth and third centuries B.C., Herophilus and Erasistratus of that school described mesenteric lymph vessels and glands. Erasistratus is quoted by Galen thus: 'For on dividing the epigastrium and along with it the peritoneum we may see arteries in the mesentery of sucking kids, full of milk' (Cruickshank, 1786).

Little progress was made in the investigation of the lymphatic system until the Renaissance when the Italian schools of medicine made very significant contributions. Eustachius, Professor of Anatomy in Rome, described the thoracic duct and its subclavian tap in the horse in 1563. The greatest landmark in the subject at that time, however, was the experimental work of Asellius (1581–1625), Professor of Anatomy and Surgery in Milan (Plate 1.1). In 1622 he demonstrated an extensive system of ramifying white vessels, 'venae albae et lacteae', in the mesentery of a dog which had been fed shortly before its death. Incising one of these with a sharp scalpel, he observed a gush of white fluid. These vessels were not readily demonstrated in a second animal which had been starved, but Asellius correctly assumed that the vessels would be best seen in the post-prandial state. He confirmed this in an experiment on a third animal. Plates 1.2 and 1.3 are taken from his thesis, *De Lactibus sive lacteis venis*, published in 1627 by two of his friends two years after his death; Plate 1.3 shows the blood and lymph vessels of the dog small intestine. Incidentally, this thesis is believed to be one of the earliest texts of medical science illustrated in colour. Asellius believed that mesenteric lymphatics ran to the liver conveying absorbed nutrients for the 'concoction' of blood in that organ. It is interesting that the part played by intestinal lymph vessels in the absorption of fat should have been responsible for such important observations as those of Erasistratus and Asellius.

An exciting phase of rapid progress in studies of the lymphatic system followed the work of Asellius. Jean Pecquet of Dieppe published his *New Anatomical Experiments* in French in 1651 and in English in 1653. In this thesis he described the communication of the 'receptaculum chyli' (cisterna chyli) with the thoracic duct and noted that intestinal lymphatics drained into the cisterna chyli (Plate 1.4). The first descriptions of human intestinal lymphatic vessels were made at this time, by Vesling of Padua (Lord, 1968) and by Pecquet. At or about the same time, other workers such as Vesalius and Jan van Horne, Professor of Anatomy at Leyden, described the human thoracic duct.

It is difficult to be certain where credit belongs for scientific discoveries in this period. Public demonstration and defence of scientific thesis was a common means of announcing one's discoveries, but communication of such information between individuals working in medical schools in different countries was erratic and printed publication of work was a slow procedure by comparison with modern standards. As a result, studies proceeded in parallel in different medical schools and the credit for an important discovery was often in question. A famous dispute of this kind arose between the Swede, Olaf Rudbeck (1630–1708) and Thomas Bartholin (1616–1680), Professor of Anatomy at Copenhagen. These two workers independently contributed a great deal to the understanding of the lymphatic system in demonstrating the ubiquitous nature of the lymph vascular system and in integrating the numerous observations which had been made by their predecessors and contemporaries.

Rudbeck was an extraordinary man who spent much of his life in Uppsala and became Rector of that University at the age of thirty-one. He was a professor in the medical faculty but also taught chemistry, astronomy, architecture and music. His interests also included scientific studies of botany and the history of Sweden and its culture. His work on the lymphatic system, performed largely between 1650 and 1653, dated from his observation of chyle flowing from the thoracic duct in the neck of a slaughtered calf. During this period he studied lymph vessels of the liver, abdominal wall and thoracic viscera, and traced the course of vessels lying on the posterior surface of the rectum to the cisterna chyli. In 1652 he gave a famous demonstration of his findings to Queen Christina of Sweden. His studies showed that the lymphatic vessels constituted a system comparable with the blood vascular system (Plate 1.5), and he was able to demonstrate that lymph in the cisterna chyli does not flow to the liver as had been supposed by Asellius. This confirmed the observations of Pecquet, and Galenical theory was seriously questioned.

To establish the direction of lymph flow, Rudbeck used occlusion techniques:

The following year, I opened the abdomen of a cat, five hours after a full meal. In order that the chyle should not run off quickly, I placed two ligatures on the lacteal vessels, one above, the other below the pancreas where the mesentery is attached to the back. Then I turned to the thorax, opened it, removing the sternum, and ligated the same duct which I had seen the previous year in the calf. After freeing the mesentery connecting the intestines with the lower lobe of the liver, I pushed aside the intestines towards the left and opened the ligature below the pancreas. As a result the chyle flowed

3

out of some small vessels into a vesicle situated between the diaphragm and the kidneys, posterior to the vena cava and the aorta, the vesicle becoming distended. I again tied this ligature and released the upper one; the result of this was that the upper chylous vessels emptied, while the lower ones became filled as far as the ligature. When, in turn, the ligature below the pancreas was loosened, the chyle immediately forced its way into the vesicle. (From a translation of Rudbeck's *Nova Exercitatio Anatomica* (1653) published by Nielsen in 1942.)

In subsequent experiments, Rudbeck established that lymph flowed from the thoracic duct to the venous circulation.

An Englishman, George Joliffe, working in Cambridge at this time, appears to have made detailed study of the lymphatic system and some of his observations may have preceded those of Bartholin and Rudbeck; his early death, however, prevented publication of his results. His work was reported in 1654 by the anatomist Glisson (Nielsen, 1942).

Several workers, including Rudbeck, Bartholin and Joliffe, had described valves in lymphatic vessels and had recognised their functional importance. The principal description of these structures, however, is to be found in the thesis of the Dutch anatomist, Frederick Ruysch, *Dilucidatio Valvularum in Vasis Lymphaticis et Lacteis* (1665) (Plate 1.6).

During this intense phase of activity in the seventeenth century, injection techniques were developed and employed by several workers including Pecquet and Ruysch, for visualising the finer lymphatic vessels. Another Dutch anatomist, Nuck, described the use of mercury for this purpose in 1696, and during the eighteenth century such techniques were widely used in painstaking topographical studies of the lymphatics of many species. The magnificent illustrations in the treatise of Mascagni, *Vasorum Lymphaticorum Corporis Humani* (1787), exemplify this type of study (Plate 1.7).

Along with increasing understanding of the anatomy of the lymphatic system came the acceptance of the concept of the absorptive function of lymphatics. Much of the work supporting this idea came from the Hunterian school of anatomy which was particularly active in lymphatic research. Associated with this group was William Hewson who published his *Experimental Inquiries* in 1774. In part II of this volume he presented an account of his investigations of lymphatic anatomy and

Plate 1.1. Portrait of Asellius. (Courtesy of the Wellcome Trustees.)

Facing p. 4

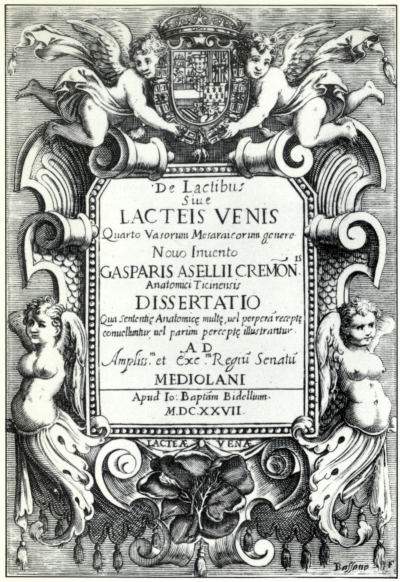

De Lactibus
Siue
LACTEIS VENIS
Quarto Vasorum Mesaraicorum genere
Nouo Inuento
GASPARIS ASELLII CREMON.is
Anatomici Ticinensis
DISSERTATIO
Qua sententię Anatomicę multę, uel perperā receptę
conuelluntur, uel parūm perceptę illustrantur.
AD
Ampliss.m et Exc.mRegiū Senatū
MEDIOLANI
Apud Io: Baptām Bidellium.
M.DC.XXVII.

LACTEÆ VENA

Plate 1.2. Title page of *De Lactibus sive lacteis venis* by Asellius, 1627. (Courtesy of the Wellcome Trustees.)

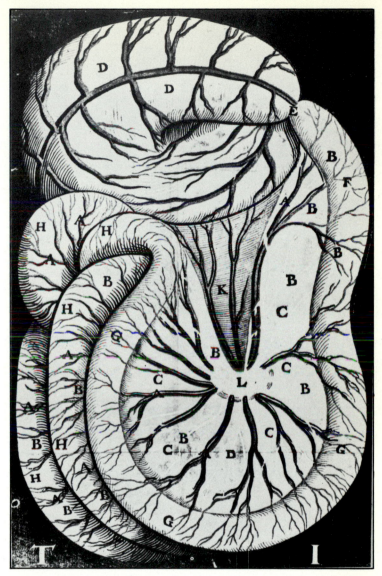

Plate 1.3. Blood and lymphatic vessels of the dog alimentary tract from *De Lactibus sive lacteis venis*, 1627. (Courtesy of the Wellcome Trustees.)

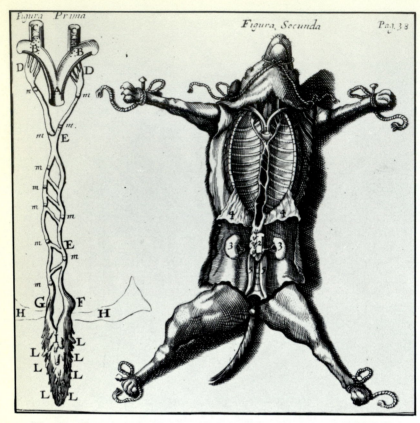

Plate 1.4. Thoracic duct and cisterna chyli from *New Anatomical Experiments* by Jean Pecquet, 1653. (Courtesy of the Wellcome Trustees.)

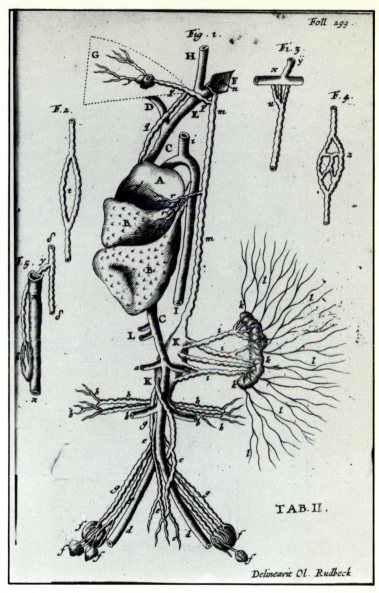

Plate 1.5. Major lymph trunks from *Nova exercitatio Anatomica* by O. Rudbeck, 1653. (Courtesy of the Wellcome Trustees.)

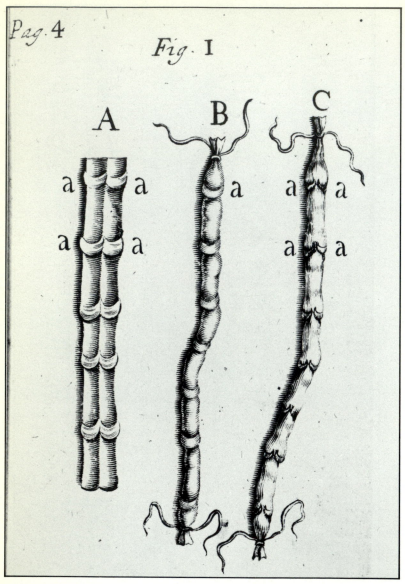

Plate 1.6. Lymphatic valves from *Dilucidatio Valvularum in Vasis Lymphaticis et Lacteis* by F. Ruysch, 1665. (Courtesy of The Wellcome Trustees.)

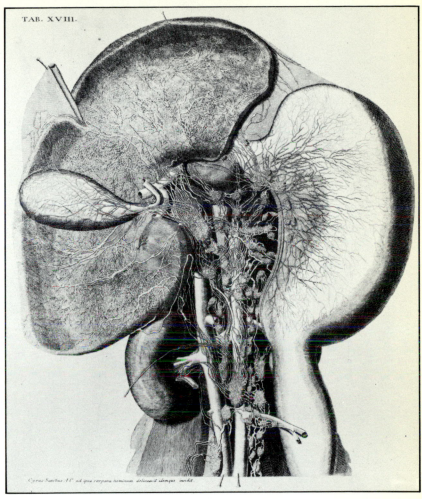

Plate 1.7. Anatomy of hepatic and gastric lymph vessels in man from *Vasorum Lymphaticorum Corporis Humani* by P. Mascagni, 1787. (Courtesy of the Wellcome Trustees.)

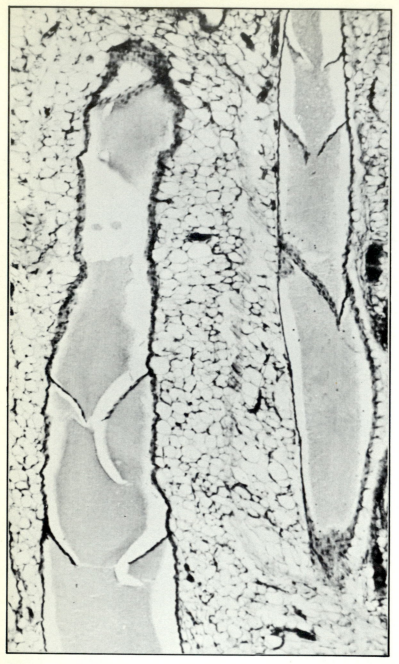

Plate 2.1. Lymphatic valves in slightly dilated lymphatics in rabbit mesentery, a low-power magnification. (Kalima, 1971.)

physiology which included a detailed account of the central lymph vessel of the intestinal villus – a structure which Lieberkühn had studied at that time with the microscope. Hewson observed motility of lymphatic vessels referring to 'peristaltic contraction of the coats' of the vessels and considered that this activity together with the transmitted pulse from neighbouring arteries would account for lymph propulsion, the valves of the lymphatic vessels contributing a directional component to the flow. Hewson's work was extended by an associate, William Cruickshank, who in 1786 made the interesting observation that certain dyes introduced into the intestinal lumen were taken up by mesenteric lymphatics.

The topography of the lymphatic system and the general principles of function of the system were now becoming clear but the origin of tissue fluid and the formation of lymph still required clarification. Ludwig (1858) proposed a theory of formation of tissue fluid based on physical factors governing fluid filtration across blood capillary walls. This theory was challenged by Heidenhain (1891) who postulated the existence of specific chemical agents which would stimulate lymph production. It was Starling whose work (1894) led to a theory which embodied the ideas of Ludwig and expanded them. His theory of fluid balance across capillary walls was based on physical factors, as was Ludwig's, but it included the important osmotic pressure component contributed by plasma proteins, to which the capillaries are relatively impermeable. This laid the foundation of modern concepts of fluid balance between blood, the interstitial space and lymph.

2 ANATOMY OF THE LYMPHATIC SYSTEM OF THE GASTRO-INTESTINAL TRACT

The gross anatomy of the principal channels of the mammalian lymphatic system is depicted in Fig. 2.1. This shows the main sites of discharge of lymph into the venous circulation at the root of the neck. The principal lymphatic trunks are found on the left side, the bulk of lymph entering the circulation through the left jugulo-subclavian tap (the point of entry of the thoracic duct into the jugular vein in Fig. 2.1). It should be emphasized that this anatomical arrangement is subject to considerable inter-species variation and to differences within individuals of a species. These variations include local failure of development of lymph trunks with formation of compensatory collateral channels and lymphatico-lymphatic anastomoses. These, together with the development of accessory lymphatico-venous anastomoses which are discussed in detail below, are mechanisms where the lymphatic system shows its great capacity for bypassing obstruction.

The lymph draining the abdominal viscera is ultimately collected in the cisterna chyli. Of the lymph flowing in this channel, a large proportion is derived from the organs of the gastro-intestinal tract, especially the liver and small intestine. The routes taken by the draining vessels from each organ vary amongst different species, and a detailed account of these routes is outside the scope of this discussion. Lymph in the lymphatic trunks draining an organ passes through regional lymph nodes; these nodes are often situated at the confluence of a group of major trunks, and while some lie in the main stream of the lymph, major lymphatic vessels frequently bypass lymph nodes, and small 'side-arm' lymph vessels connect these nodes with the large trunk. Fig. 2.2 shows the anatomical arrangement of the lymphatic vessels draining the abdominal viscera of the rat.

6

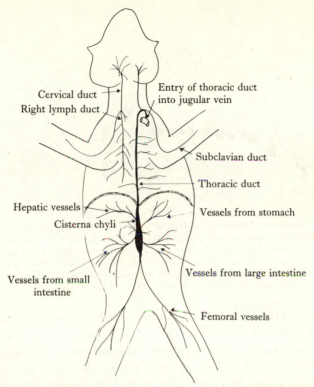

Fig. 2.1. Schematic representation of the principal lymphatic trunks in mammals. (Redrawn from Yoffey & Courtice, 1956.)

The histological arrangement of a lymph node is well described in Rhodin (1974) and will not be discussed here. In addition to lymph nodes, which act as filters for lymph passing through the larger lymph vessels, abundant lymphatic tissue is also found distributed throughout the length of the gastro-intestinal tract close to the epithelial surfaces. In some areas, lymphatic tissue forms large aggregates such as the pharyngeal tonsils and Peyer's patches of the ileum. The tissue in these subepithelial areas is not encapsulated and so interstitial fluid in the lamina propria of the small intestine, for example, can percolate through this lymphatic tissue, which acts as a filter for the tissue fluid, prior to its uptake by either lymph or blood vessels.

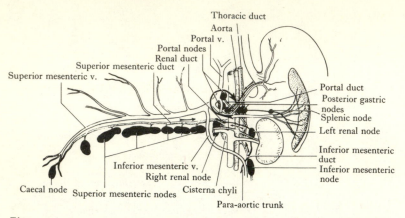

Fig. 2.2. Anatomy of the major abdominal lymphatic vessels of the rat. Lymphatics of the abdominal viscera drain into three major lymph ducts which enter the cisterna chyli. The portal duct drains the nodes associated with the upper abdominal organs, the superior mesenteric duct drains the small intestine and the ascending and transverse colon, while the inferior mesenteric duct drains the descending and sigmoid colon. Arrows demonstrate the direction of lymph flow. (Tilney, 1971.)

The lymphoid tissue of Peyer's patches is usually situated on the antimesenteric border of the intestine. Each patch typically comprises forty to fifty nodules which are separated from the gut lumen by a layer of epithelial cells and from the serosa by a thin layer of vascularised connective tissue. The patches appear to have their own blood supply. Follicles are separated from each other by intervening villi. Three layers are distinguished in each nodule: an outer layer close to the serosa comprising large lymphocytes and containing some germinal centres, a middle zone consisting of small lymphocytes and a zone close to the epithelial cell surface containing a mixed cell population (Faulk *et al.*, 1971).

Some detailed consideration of the anatomy of lymph vessels in general and of the lymphatic vasculature of the individual organs comprising the gastro-intestinal tract is valuable for an understanding of physiological function of this system.

The lymphatic vessels

While the major lymphatic vessels are relatively easy to identify, the recognition of the fine terminal vessels of the system
is a formidable problem since these vessels have a tenuous wall
and tend to collapse in histological preparation. Lymphatic
vessels are lined by a flattened endothelium with an incomplete
basement membrane. This endothelium is the only true component of the wall of fine lymphatic capillaries which terminate
as blind-ended tubes. Fine strands of connective tissue are
closely associated with terminal lymph vessels, and these
may have some supportive function (Pullinger & Florey, 1935).
Vessels of greater diameter, transporting vessels, acquire
smooth muscle and elastic elements in a connective tissue layer.
Carleton & Florey (1927) found that in the cat mesentery,
medium-sized lymph vessels (of 100–200 μm diameter) have
rather sparse smooth muscle elements in their walls, while
smaller lymphatic vessels in this species have none. Rats, mice
and guinea pigs, however, have abundant smooth muscle in
the walls of medium-sized mesenteric lymph vessels, and in
smaller vessels of approximately 50 μm diameter, recognisable
smooth muscle elements are present. The contractile activity
of lymphatic vessels is probably related to the amounts of
smooth muscle in the vessel, since spontaneous activity of
lymphatics has, in general, been most easily demonstrated in
such species as rats and guinea pigs which have a prominent
component of smooth muscle in the vessel wall (see Chapter
3). Large lymph vessels have a generous supply of nutritive
blood vessels in their adventitia. These vasa vasorum were first
described by Cruickshank (1786) and subsequently by Dogiel
(1883). In addition to supplying nutrients to lymphatic smooth
muscle, these vessels may modify lymph composition (see
Chapter 3).

The valves of lymphatic vessels which are prominent in
mammals though absent in fish, reptiles and amphibians, are
composed of single or double cusps (Plate 2.1). In general, they
are seen in vessels larger than the capillaries and, together with
lymphatic smooth muscle, assist in the propulsion of lymph.
Their presence, together with ampullary dilations between

9

valves, confers a beaded appearance on larger lymph vessels. Thus lymphatic vessels can be considered as chains of units, each segment bounded by valves, and the term 'lymph-angion' has been applied to these units (Mislin, 1961). The smooth muscle of lymphatic vessels is characteristically found in the regions between valves, while at the origin of the valves, the lymphatic wall has little or no smooth muscle (Schipp, 1967).

Prelymphatics
It has become clear recently that the interstitial space of tissues is not merely an amorphous mucopolysaccharide gel, but is permeated by channels through which fluid can move with little impedance, unlike passage through the gel itself which would offer resistance to flow. The term 'prelymphatics' has been coined to describe a group of such non-endothelialised channels, which have been demonstrated by morphological techniques. Prelymphatics frequently end at junctions between endothelial cells of lymph capillaries (Collan & Kalima, 1974; Casley-Smith, 1976).

Lymphatico-venous anastomoses

The question as to whether lymphatico-venous anastomoses exist at sites other than the root of the neck and, if so, whether they are of functional importance, is an important problem of particular consequence to the investigator who tries to determine the relative importance of the lymphatic and portal venous routes in the transport of absorbed material.

Since the sixteenth century when Eustachius described the jugulo-subclavian tap, the possible existence of other lymphatico-venous communications has been discussed. The close embryological origin of veins and lymphatics postulated by Sabin (1916) might account for the presence of such communications in adult life. This concept of the embryological origin of lymphatics has, however, been challenged by Kampmeier (1960, 1969) who considers that lymphatic endothelium is derived from mesenchymal elements and not from a 'budding off' of venous endothelium. There is an extensive and confusing literature about the existence of lymphatico-venous anastomoses. One can recognise several factors which have led to the confusion: if a *potential* channel exists which is not ordinarily patent, there is a possibility that this will open as a result of the investigative technique, that is, pressure produced by injection of dyes or contrast media, or as a result of lymphatic obstruction. Again, physiological circumstances in which intra-lymphatic pressure rises, such as occur during lymphatic transport of large volumes of fluid, might cause these channels to open. Species, strain and individual differences which undoubtedly exist result in conflicting reports on this subject, and minor developmental anomalies may account for a proportion of lymphatico-venous communications which are demonstrated in man and animals. The criteria used by various investigators for demonstrating the existence of these channels vary greatly. Anatomical and radiological studies demonstrate large gauge anastomoses, while the presence of microscopic communications is inferred from studies of the passage of particles, fluids and gases from one channel to the other.

From a physiological point of view, an important question is whether a regular functional lymphatico-venous tap exists, for example, between the major intestinal lymph trunks and the hepatic portal or systemic venous system. Two studies early in this century suggested that numerous lymphatico-venous anastomoses exist in South American monkeys (Silvester, 1911) and in the wild rat (Job, 1918). Silvester showed, by means of injections with coloured gelatin, that in the monkeys he studied most abdominal lymph entered the renal vein or the inferior vena cava. These animals do not have a cisterna chyli or thoracic duct; neither does the sloth, *Bradypus tridactylus*, which also has numerous abdominal lymphatico-venous com-munications (Azzali & Didio, 1965). Roddenberry & Allen (1967) have extended the study of lymphatico-venous anasto-moses in South American monkeys, identifying them by pul-satile reflux of venous blood into the first segment of the lymphatic. They found that lymph continued to flow into the vein until the opposing venous pressure reached 10 mmHg. It was observed that the first lymphatic valve would withstand a pressure of about 45 mmHg when saline was infused into the inferior vena cava. These authors found no evidence of a sphincter mechanism at the lymphatico-venous ostium.

While there is no doubt about the presence of lymphatico-venous communications in New World monkeys, there is no general agreement regarding their existence in the rat. Job (1918), injecting India Ink into the lumbar lymph chain of wild rats, found that considerable amounts entered the vena cava at the level of the renal veins. This type of communication was more frequently demonstrated in pregnant rats. Threefoot, Kent & Hatchett (1963), using plastic corrosion models of the lymphatic system of the laboratory rat, have demonstrated lymphatico-venous communications only when the system was stressed by cisterna chyli ligation. Such communications might account for the observations of Paldino & Hyman (1964) who found in laboratory rats that lymph flow in the main intestinal lymphatic vessel cannulated close to the caecum is almost as great as that in the thoracic duct but that occlusion of the ileo-colic vessel diminishes the thoracic duct flow to only a very slight degree.

Roddenberry & Allen (1967) were unable to identify lymphatico-venous communications in Sprague–Dawley rats by the criteria which they had used in squirrel monkeys. In view of the mass of physiological studies of intestinal lymphatics in the rat, it is important to come to a firm decision about the presence of these communications in this species.

Under normal circumstances, they are probably uncommon but they must be taken into account in determining the relative importance of the portal venous and lymphatic routes in the transport of any absorbed substance. For example, a substance such as cholesterol which is transported from the intestine by lymph may appear in small amounts in portal venous blood and appear in systemic blood despite diversion of intestinal lymph. It is clear that partial obstruction of an experimental lymph fistula may be sufficient to establish lymphatico-venous diversion, thus making it difficult to assess whether the portal venous route has a normal minor role in transport of such a substance.

In sheep, Heath (1964) has concluded that communications between intestinal lymphatics and the venous system are not

common if, indeed, they occur at all. Albumin labelled with
[131]I introduced into a mesenteric lymph vessel was almost
entirely recovered in the thoracic duct. Immediately after
occlusion of the thoracic duct, labelled albumin given by the
lymphatic route appeared very slowly in the circulation but five
days later, the rapid appearance of the tracer in the blood after
intra-lymphatic injection indicated the establishment of a
lymphatico-venous channel (Fig. 2.3). Over several weeks,
radiologically demonstrable lymphatico-venous anastomoses
appeared below the obstruction.

Obstruction of the main lymph trunks is followed by the
establishment of lymphatico-venous anastomoses in cats (Lee,
1922; Carlsten & Olin, 1952) and in dogs (Blalock, Robinson,
Cunningham & Gray, 1937). Freeman (1942) and Glenn *et al.*
(1949) have described lymphatico-venous communications in

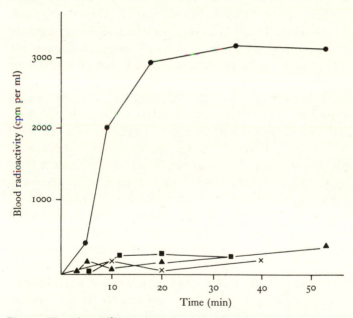

Fig. 2.3. Transfer of ^{131}I-labelled albumin from intestinal lymph to systemic blood
in the sheep at intervals after thoracic duct occlusion. ($\times$, immediately after occlusion;
▲, one day after occlusion; ■, two days after occlusion; ●, five days after occlusion.)
Ordinate represents level of radioactivity in samples of jugular vein blood. Abscissa
represents interval after injection of labelled albumin into a mesenteric lymphatic.
(Heath, 1964.)

the dog. Neyazaki, Kupic, Marshall & Abrams (1965), in a radiological study, found that ligation of the thoracic duct of dogs resulted in distension of obstructed lymph vessels which persisted for at least three weeks before lymphatico-venous channels opened.

Pressman & Simon (1961), Pressman, Simon, Hand & Miller (1962) and Pressman, Burtz & Shaffer (1964), consider that microscopic lymphatico-venous communications are present in lymph nodes, finding that fluids and air injected into lymph nodes of dogs are able to pass into the veins draining the nodes. These authors further observed that bacteria and particles of up to 25 μm diameter were able to pass by this route. Neither the trauma of injection nor excessive injection pressure was felt to contribute to this effect. Nevertheless, in the *unobstructed* lymphatic system, albumin infused centrally into a leg lymphatic of the dog is almost totally recovered in the thoracic duct. Fig. 2.4 (Patterson, Ballard, Wasserman & Mayerson, 1958) demonstrates this. There is a considerable weight of evidence that small amounts of water and small molecules leave the lymph to be taken up by the bloodstream in lymph nodes (see Chapter 3).

In man, lymphatico-venous anastomoses of large gauge have been frequently demonstrated radiologically. Pick, Anson & Burnett (1964) described such communication at the level of the renal vein. Often, there is an associated element of lymphatic obstruction. For example, Retik, Perlmutter & Harrison (1965) reported communication between lymphatics and the portal vein in carcinomatous obstruction of retroperitoneal lymphatics. The frequency of pulmonary oil embolism following lymphangiography, especially when lymphatic obstruction is present, suggests the presence of lymphatico-venous channels.

Threefoot *et al.* (1963) and Threefoot & Kossover (1966) have studied this problem in man *post mortem* by an interesting combined technique. A long gauze roll was inserted into the inferior vena cava and radio-opaque radioactive material was injected into abdominal lymphatics. X-rays were taken, and the gauze roll removed and scanned for radioactivity. The position of the radioactivity on the gauze was correlated with the

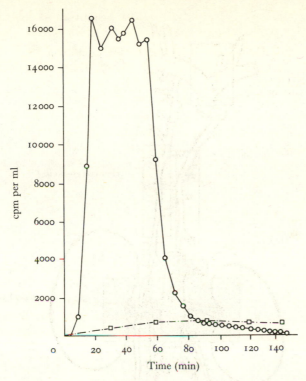

Fig. 2.4. Appearance of ^{131}I-labelled albumin in lymph and plasma following its infusion centrally into a leg lymphatic at zero time. Albumin infusion stopped after 50 min and 0.9% saline infusion started and continued for the next 100 min. (Patterson *et al.*, 1958.)

lymphangiographic demonstration of communications between lymphatics and veins.

From the available evidence about the occurrence of these communications, no simple conclusion can be drawn. Rather, one is forced to conclude that these communications exist in many species, becoming functionally important only when stressed by rising lymphatic pressure. Fig. 2.5 illustrates the sites of communication in mammals and their frequency. Clearly, potential channels might open under pressure in a simple mechanical way; but Threefoot *et al.* (1963), having observed that systemic administration of hexamethonium facilitated their demonstration in rats, have made the interest-

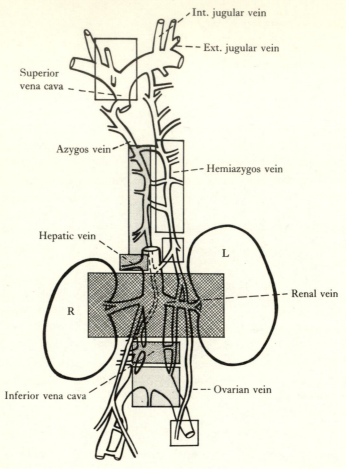

Fig. 2.5. Sites of lymphatico-venous communications and their frequency in mammals. The degree of frequency is indicated by the density of shading. (Threefoot, 1968.)

ing suggestion that an autonomic reflex triggered by raising intralymphatic pressure may mediate the opening of these shunts.

Alimentary tract lymph vessels

Lymphatics of the oesophagus and stomach

The alimentary tract is generously supplied with lymphatic vessels. Layers of lymph vessels are found in the tissues of each part of the gut. As a general rule, blood capillaries lie in a more superficial position in relation to the mucosal epithelium than terminal lymph vessels. In the oesophagus, stomach and intestine there is a plexus of lymph vessels present in mucosal, submucosal and muscular layers and communication from one plexus to the next is through short vessels running perpendicularly to the gut axis. Valves in the vessels of the submucosa and muscular layers allow lymph to be directed towards the collecting vessels draining the organ. The average diameter of lymph vessels increases progressively from the mucosa to the external surface of the organ. Mall (1896) observed that when he stripped the mucosa of dog stomach from the muscularis mucosae in a preparation which had been injected with dye, a mucosal lymphatic network lay close to the plane of cleavage. Draining into this network were blind-ended terminal lymph vessels passing among the basal parts of the gastric glands. Rényi-Vámos & Szinay (1954) have emphasised the close proximity of these fine lymphatic capillaries to the epithelium of the gastric glands.

Functional communication between the gastric and duodenal lymph vessels through anastomotic connections in the wall of the gut has been the subject of debate though such communications have been demonstrated in man by Zhdanov (1953). In dogs, however, Nusbaum, Baum, Rajatapati & Blakemore (1967) failed to demonstrate these. In the dog, a network of fine lymphatic vessels is present in the greater omentum. These vessels were studied by Nylander & Tjernberg (1969) using Thorotrast and Patent Blue V injected into the margin of the omentum. Flow in this system is slow. It was observed that these vessels link up with gastric lymphatics.

Lymphatics of the small intestine, the lacteals

The specialisation of the mucosal surface of the small intestine, which greatly enhances its absorptive capacity, is the villus. Intestinal villi are approximately 1 mm in length in man and are present in the entire small intestinal mucosa. Their density and height are greatest in the duodenum, decreasing towards the ileum. Finger-like villi have been found in man, dog, cat and several other species; while in rabbits, rats, mice and squirrels, flat and leaf-shaped villi are present. Finger-shaped villi have a single central lymphatic vessel while leaf-like villi may have several lymphatic vessels (Ranvier, 1896).

Since Lieberkühn's microscopic studies of the central lymphatic vessel of the villus in the eighteenth century, numerous workers have examined this vessel; its delicate structure renders it difficult to visualise in tissues prepared by conventional histological techniques. Indeed, there was still controversy regarding its existence some twenty years ago. Techniques such as those employed by Papp *et al.* (1962; see below) have allowed the unequivocal demonstration of this vessel in the villi of the small intestine and, recently, Lee (1969b) has succeeded in cannulating this vessel in villi of dog small intestine *in vitro* (see Chapter 4).

The anatomical arrangement of the blood and lymphatic vessels in the mucosa of the small intestine is basically similar to that of other parts of the gut. Fig. 2.6 illustrates the disposition of blood and lymphatic vessels in a villus of the small intestine. As in other parts of the gastro-intestinal tract, a network of capillary vessels lies more superficially in the mucosa than the terminal lymph vessels. The central lymphatic, or lacteal, of the intestinal villus can be regarded as an elongated form of the blind-ended vessels which are seen to run from close to the glands of gastric or colonic mucosa towards a plexus of lymph vessels on the muscularis mucosae. As in the stomach, so in the submucosa of the small intestine, a plexus of vessels is found which Vajda & Tömböl (1965) consider as having a reservoir function. Larger vessels draining this plexus form networks of vessels in the muscular and serosal layers which give rise to the principal efferent trunks from the organ.

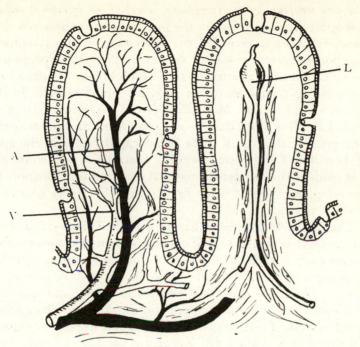

Fig. 2.6. The vessels of the intestinal villus. A, the central artery, breaks up as a network of capillaries lying just below the basement membrane of the epithelium. V is the central vein and L the lacteal. Some smooth muscle elements lie in close association with the lymphatic vessels. (Yoffey & Courtice, 1956.)

The innervation of mesenteric lymph vessels. Vajda (1966) has studied the innervation of the larger lymph vessels and lymph capillaries in the mesentery of the small intestine in dogs and cats using a silver stain. The large vessels are supplied by two types of nerve, myelinated fibres ramifying on the adventitial surface of the vessel considered to have a sensory function, and unmyelinated nerves penetrating into the muscular layer. The latter appear in greatest density in the region of attachment of valves. These fibres are considered to be the motor supply to the smooth muscle though smooth muscle is relatively sparse at valve attachments. The fine lymphatic capillaries, which consist of a single layer of endothelium, do not have a true nerve supply, but free and encapsulated nerve endings lie in particularly close relation to the lymph capillaries.

The fine structure of the lymphatic vessel of the intestinal villus
It is appropriate to consider the fine structure of this vessel in some detail as it provides the first stage in the lymphatic transport of certain absorbed nutrients. The lamina propria of the villus contains a central artery which breaks up into a very dense subepithelial plexus of capillary blood vessels. In the jejunal villi of the rat, Mohiuddin (1966) has estimated that no epithelial cell is more than 30 μm away from the bloodstream. The greatest density of these blood capillaries is at the apex of the villus. Lymphatic vessels lie centrally in the villus, and it is evident that absorbed material destined for transport by the lymphatic vessels must first bypass this dense plexus of blood vessels.

The central lymph vessel has been studied in several species. Its structure is similar to that of terminal lymph vessels in other tissues such as those of the ears of mice and guinea pigs, the diaphragm of mice and the colon of rats (Casley-Smith & Florey, 1961). The walls of these vessels were found to consist of a single layer of thin endothelium and resemble capillary blood vessels. There are, however, certain important characteristics which distinguish lymphatic capillaries from blood capillaries: fenestrations of 20 to 50 nm, which are typically seen in the attenuated parts of capillary endothelial cells in certain sites such as intestine, are not found in lymphatic endothelium; gaps in the intercellular junctions are occasionally seen between endothelial cells of lymphatics and, in general, adhesion devices between endothelial cells are less frequently found in lymphatic walls than in capillary walls; the basement membrane, which is usually complete around capillary endothelium, is frequently deficient in relation to lymphatic endothelium (Table 2.1; Plate 2.2).

With respect to those features, the lymph vessel of the intestinal villus is similar to other peripheral lymph vessels. In the lymphatics of rat intestinal villi, Palay & Karlin (1959*a*, *b*) observed the deficient basement membrane and occasional intercellular gaps. They also noted the presence of strands of smooth muscle associated with the lymphatic endothelium which were not strictly an integral part of the wall of the vessels. In this same species, Casley-Smith (1962) identified

TABLE 2.1. *Comparison of electron microscopic appearances of the blood and lymphatic capillaries of the intestinal villus of the cat.* (Modified from Papp et al., 1962.)

	Substance in lumen	Shape	Basement membrane	Fenestration	Endothelial front	Associated cells
Terminal lymphatic vessel	Light	Irregular	Absent or fragmentary	Absent	Very irregular (lamelliform projections)	None
Blood capillary	Darker	Round	Present	Present	Less irregular	Pericytes

chylomicra and lipoproteins passing through open junctions of lymphatic endothelium in jejunal villi. Mouse intestinal lymphatics also have an incomplete basement membrane (Weiss, 1955; Deane, 1964). In bats, Ottaviani & Azzali (1969) have observed that deficiencies of the basement membrane of the central lymph capillary of the intestinal villi are most marked towards the apex of the villus. The fine structure of the central lymph vessel of the intestinal villi of the dog and cat is similar to that of other species (Vajda & Tömböl, 1965).

Papp and his co-workers (1962) have examined the ultra-structure of the lacteal in the cat; by injecting 2–3 ml of serum into the muscular layers of the intestinal wall, they were able to distend the vessel artificially. The lumen of the vessel was often seen to be stellate, and there were frequent convolutions and folds in the endothelium. Thin, flat processes were noted extending both towards the luminal and abluminal surfaces of the endothelial cells. In this study, intercellular gaps such as have been described in other species were not seen. Intracellular organelles of the endothelium included a nucleus, a well-developed Golgi apparatus in the perinuclear region, elements of rough and smooth endoplasmic reticulum and some mitochondria and microtubules. The lymphatic endothelial cells were also noted to include (in the cytoplasm) several vacuoles of diameter approximately 50 nm, often in close relation to the cell membranes; these were similar to pinocytotic vesicles of blood capillaries and have been termed micropinosomes.

In addition to these, Casley-Smith (1967a) has described three other types of intracellular vesicle in lymphatic endothelium. One group, whose diameter ranges between 100 and 5000 nm, is thought to be produced by the fusion of several smaller vesicles, and these vesicles may persist for long periods in the cytoplasm. A second group also comprises large vesicles which may offer a route for transcellular transport of chylomicra (Casley-Smith, 1962), while a third group of large vesicles have associated filaments on their exterior and bristles on their interior surface. It has been suggested that this third group may selectively take up and transport proteins.

In order to study the morphological characteristics of abnormal intestinal lymph vessels in man, Dobbins (1966a) examined the lymphatics of normal small intestinal biopsy specimens by light and electron microscopy. The structure of these vessels in the human is closely similar to that of other species. Submucosal lymph vessels were readily seen on light microscopy due, presumably, to associated supporting connective tissue, but the villus vessel was hard to see in these biopsies, a difficulty encountered in another study of human biopsy specimens (Bank, Fisher, Marks & Groll, 1967). Dobbins found an incomplete basement membrane and no fenestrae in human lymphatic endothelium. Other features which distinguished blood and lymphatic capillaries in this study were the absence of associated pericytes in the lymphatic wall and frequent intercellular gaps in the lymphatic endothelium. The intestinal blood capillaries, on the other hand, had well defined adhesion plates at endothelial cell junctions. Subsequently, Dobbins & Rollins (1970) described the morphological appearances of the lacteals of mouse and guinea pig intestine, finding intercellular gaps to occur infrequently. It is possible that these are indeed less common than had previously been supposed and adhesion devices, maculae adherentes and a few maculae occludentes, are to be found between lymphatic endothelial cells. Careful histological preparation certainly reduces the chances of creating artefactual appearances of intercellular gaps. Nevertheless, the balance of opinion favours the idea that intercellular gaps do open and close and afford an important route for fluid and solutes and particles into the lymphatic capillaries, notably in the diaphragm.

The presence of smooth muscle and connective tissue elements closely associated with the lymphatic endothelium raises the interesting possibility that abluminal projections from lymphatic endothelium serve to anchor these cells to the muscle and connective tissue and thereby offer support and a possible contractile element to the central lymphatic. In 1935, Pullinger & Florey showed that collagenous fibres attached to lymphatic endothelial cells tighten and pull lymphatic vessels open when tissues become oedematous. In rats, Collan and Kalima (1970, 1974) found in the lamina propria of the intestinal villus a network of collagen fibres attached to the endothelium of the lymph vessel and to various other elements such as fibroblasts,

23

muscle cells, nerve fibres and the blood capillary network. These authors also described the valve-like appearance of some endothelial cell junctions in the lymph vessel and proposed that when fluid fills the interstitial spaces of the villus, the valve-like structures open as the collagen fibres attached to the abluminal surface of overlapping endothelial cells tighten, thus creating a channel into the lymph vessel (Fig. 2.7). These endothelial cell junctions are postulated to act as simple flap valves which subsequently close when the intraluminal pressure equals or exceeds the interstitial pressure. The presence of fine filaments of approximately 6 nm diameter which link lymphatic endothelium with surrounding connective tissue has also been described with respect to other lymph vessels such as those of the guinea pig ear (Leak & Burke, 1966, 1968). Collagenous bundles have been observed in close relationship to lymphatics in the portal tracts of the liver, and it is likely that this is a structural support found in relation to all fine lymph vessels which functions in the manner described above. However Dobbins & Rollins (1970), in a study of mice and guinea pigs, failed to demonstrate these anchoring filaments around the central lacteals of intestinal villi, though they were seen in relation to

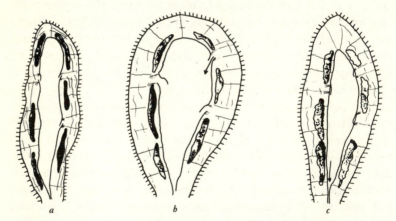

Fig. 2.7. Diagrammatic representation of a hypothetical model for the function of the central lacteal of the intestinal villus. Collagenous fibres attached to lymphatic endothelium tauten (*b*) when fluid flows into the lamina propria. The villus rounds out and the fibres open valve-like formations between endothelial cells allowing fluid to enter the lymphatic lumen. Subsequently the contraction of smooth muscle elements in the villi force this fluid toward the submucosal lymphatics. (Collan & Kalima, 1970.)

submucosal lymphatics. It is very likely that the longitudinally disposed smooth muscle cells in intestinal villi are responsible for the 'villus pump' action which is probably of major importance in transport of lymph in lacteals.

As far as the central lymphatic of the intestinal villus is concerned, it is becoming apparent that associated structural and contractile elements must be taken into account in considering its functional properties.

Lymphatics of the large intestine
Numerous lymph vessels are found in the wall of the large intestine. In the mucosa of the colon, blood capillaries lie immediately below the epithelium, while the lymph capillary network lies arranged horizontally and rather more deeply, close to the muscularis mucosae and the base of the intestinal glands (Kamei, 1969). Blind lymphatic capillary projections arise from the network among the basal parts of the glands. Lymphatic capillaries in this area are of much greater calibre than the blood capillaries, and vessels draining this network are seen in the submucosa running close to blood vessels. In dog and cat colon, large lymph vessels equipped with valves are seen passing through the muscle layer (Vajda & Leranth, 1967). In this region, these vessels were noted to show ampullary dilatations presumably of intervalvular segments, and these authors considered the lymphatic network of the large intestine may have a high storage capacity.

The lymphatic vessels of the dog colon are segmentally arranged. When the lymphatics draining a segment are surgically ablated, oedema develops in the bowel wall but persists for only about three weeks, after which an effective collateral circulation is established (Sterns & Vaughan, 1970). On the other hand, Nesis & Sterns (1973), measuring the removal of ^{131}I-labelled contrast medium given as a subserosal injection in the colon, reported that an effective collateral circulation was not re-established in four weeks after removal of colonic segmental lymph vessels and nodes.

Lymphatics of digestive glands

The salivary glands

In studies of man, ape, dog, rabbit and guinea pig, Klein (1882) identified lymphatic vessels in connective tissue tracts in salivary glands. Two groups of vessels were found, both containing valves: one type of vessel was found in relation to salivary ducts and the other type close to blood vessels. The lymphatics in this latter group often surrounded blood vessels as a plexus or were seen as large single vessels almost completely enclosing the blood vessels. Papp & Fodor (1957) have made similar observations and have stressed the proximity of terminal lymph capillaries to the basement membrane of the acini. They have also described a system of lymphatic vessels in the capsule of the glands.

The pancreas

Bartels (1909) observed that lymph vessels of the pancreas form a widespread network in the tissue between lobules of glandular cells and considered that no lymphatic vessels penetrated the Islets of Langerhans. These findings have been confirmed by modern work (Rusznyak, Földi & Szabo, 1967).

Földi, Kepes & Szabo (1954) have reported, in a study of a few dogs, that they could demonstrate, in addition to the larger valved lymphatic vessels which accompanied blood vessels in the connective tissue septae, fine lymphatic capillaries passing through lobules to lie in close relationship with the basement membrane of the acinar cells. There is, however, some doubt about the degree to which finer lymph vessels come in contact with the secretory epithelium of the exocrine pancreas. Using dye injections, the lymphatics of the pancreas of dogs, horses and rats were only seen in interlobular tissue and did not appear to penetrate the lobules (Grau & Taher, 1965); as in Bartels's studies, no lymph vessels were found to penetrate the Islets of Langerhans. Godart (1965) has examined the pancreas of anaesthetised rats using stereoscopic microscopy with transillumination. Following injection of a mixture of Patent Blue dye and India Ink into the pancreatic duct it was noted that while the carbon granules of the ink were retained within the duct system for some time, the dye passed between the epithelial cells into the interstitial tissue and in 10–15 min filled local lymph vessels. The vessels, which were often associated with blood vessels, ran between the lobules but never penetrated them. In human pancreas injected with Patent Blue at laparotomy, a fine network of subcapsular interlacing lymph channels has been observed. These channels were absent in the pancreas which had been the seat of repeated episodes of acute pancreatitis, and it has been suggested that lymphatic blockage might contribute to the pathogenesis of chronic pancreatitis (Reynolds, 1970; see Chapter 5).

The liver

A substantial contribution to lymph flow in the cisterna chyli is made by the liver. Liver lymph has a high protein content, and it has been proposed that some newly synthesised plasma protein may be delivered to the circulation by this route, though the proportion delivered this way is probably very small (see Chapter 8); almost all the protein of liver lymph is derived from the blood by filtration in the hepatic sinusoids. The highly permeable nature of the sinusoidal endothelial wall is largely responsible for the high content of plasma protein in liver lymph.

Three systems of liver lymph vessels are described associated with portal tracts, hepatic veins and liver capsule (Comparini, 1969). The first system consists of initial lymph capillaries which arise in the interlobular connective tissue. Delicate valves are present in fine lymph vessels in this area. It is difficult to establish whether these fine portal lymphatics lie in relation to artery or vein, but in the large portal tracts it becomes clear that the lymph vessels lie in close relationship with the branches of the hepatic artery. There is no evidence that the portal tract system or any other system of lymph vessels takes origin from within the hepatic lobule.

A second system of intrahepatic lymph vessels is found lying close to the tributaries of the hepatic veins. These vessels are few and less conspicuous than the portal group. Lymphatic vessels do not appear to be present within the lobules. The lymph which these two systems drains is probably derived from the space of Disse between sinusoidal endothelium and hepatic parenchyma.

A third extremely complex network of lymphatic vessels is present in the fibrous capsule of the liver. In this region, it is possible to distinguish three layers of vessels. Anastomoses have been thought to exist between this capsule system and the intrahepatic lymphatic systems, but recent studies by Szabo and his colleagues (1975*b*), which examined lymph flow from the capsular region in response to occlusion of hilar lymph vessels which drain deeper parenchymal tissue, have suggested that, in dogs at least, these anastomoses are not of great functional importance. This problem is more than academic in the light of current views on the pathogenesis of ascites (see Chapter 8).

In dogs, two major hilar lymph systems have been identified: a principal system which drains the right lobe predominantly and an accessory system draining mainly the left lobe (Ritchie, Grindlay & Bollman, 1959). These authors also estimated that the hilar route accounted for about 80% of lymph draining from the liver, while 20% passed by the lymphatic vessels related to the hepatic veins.

The gall bladder

The arrangement of lymphatic vessels in the gall bladder wall was described in detail in 1883 by Dogiel. There are three main groups of vessels, viz. mucosal, muscular and subserous. The mucosal group forms an intricate irregular network; these vessels have no valves. They are connected by slender straight communicating vessels with a group of larger vessels of irregular contour, some of which have valves; these vessels ramify in the muscle layer of the gall bladder, and this group anastomoses through short thick trunks with a superficial subserosal plexus of larger valved vessels which form a network over the fundus of the gall bladder and drain into straighter vessels close to the cystic duct.

The finding, in cats, of high flow rates of gall bladder lymph which had a protein content closely similar to that of liver lymph (McCarrell, Thayer & Drinker, 1941) suggested strongly that there are anastomotic connections between liver and gall bladder lymphatics. Injection of India Ink into the liver parenchyma resulted in the appearance of the ink in the gall bladder lymphatics. Earlier observations in dogs (Sudler, 1901) had suggested that such communications might exist, though Winkenwerder (1927) failed to demonstrate them in cats.

If these communications exist, it means that true gall bladder lymph could only be obtained by some means which would obliterate the anastomotic vessels. The high protein content of gall bladder lymph observed by McCarrell *et al.* (1941) is likely to be due to a lymph contribution from the liver since it is difficult to explain why lymph from the gall bladder should have a particularly high protein concentration, whereas in the liver, the peculiar micro-anatomical arrangement of blood vessels, interstitial space and lymph vessels offers a ready explanation for the high protein content of liver lymph. Biliary tract lymphatics also appear to have functional communications with pancreatic lymph vessels in dogs. The vessels from the gall bladder and pancreas converge on a group of paraduodenal lymph nodes prior to entering the cisterna chyli. Weiner *et al.* (1970) have demonstrated that India Ink particles infused into lymph vessels of either organ will pass retrogradely to the other organ (see also Chapter 5).

The diaphragm. Although these vessels are not strictly gastro-intestinal lymphatics, they play a major role in fluid, solute and particle absorption from the peritoneal cavity. The importance of this generally relates to pathological changes in the alimentary tract and liver (see Chapters 5 and 8).

Two plexuses of diaphragmatic lymphatics are found, one on the peritoneal surface and the other on the pleural surface. These are linked by communicating vessels. Diaphragmatic lymphatics drain principally to vessels running in association with the internal mammary blood vessels to empty principally into the right lymph duct.

Material passing from peritoneal cavity to diaphragmatic lymphatics must negotiate the mesothelial layer of diaphragmatic peritoneum, and the connective tissue lying between this and the lymphatic endothelium. Most evidence suggests that fluid and particles pass through intercellular gaps in both mesothelium and lymphatic endothelium. Sac-like lymphatic lacunae are seen in this area which appear to comprise dilated lymphatic anastomotic vessels joining lymph vessels which run in parallel with muscle fibres. It is probable that rhythmic diaphragmatic contraction and relaxation are responsible for opening wide intercellular gaps and encouraging the uptake of fluid and solutes. For a detailed discussion of this subject, the reader is referred to the excellent account in Yoffey & Courtice (1970).

Summary

The tissues of the alimentary canal and its related digestive glands have an abundant supply of lymph vessels. The initial lymphatic radicles lie in close relation to absorptive and secretory epithelium and offer a route for drainage of absorbed material or interstitial accumulations of metabolic and secretory products of these cells together with fluid and protein derived from the blood vascular system, and can, therefore, be expected to play an important part in the normal function of the digestive tract.

3 GENERAL PHYSIOLOGICAL CONSIDERATIONS

This chapter discusses general aspects of the physiology of lymph formation and flow with particular attention to the lymph of the gastro-intestinal tract. Liver lymph is considered separately in Chapter 8.

A physician's explanation for a layman of the formation of a capillary filtrate and the circulation of interstitial fluid and lymph can be found in Thomas Mann's novel *The Magic Mountain* (1924):

The lymph is the most refined, the most rarefied, the most intimate of the body juices...it is the lymph that is juice of juices, the very essence...blood milk, crême de la crême: as a matter of fact, after a fatty diet it does look like milk...the blood...did not come into immediate contact with the body cells. What happened was that the pressure at which it was pumped caused a milky extract of it to sweat through the walls of the blood vessels, and so into the tissues, so that it filled every tiny interstice and cranny, and caused the elastic cell tissue to distend. This distension of the tissues, or *turgor*, pressed the lymph, after it had nicely swilled out the cells and exchanged matter with them, into the *vasa lymphatica*, the lymphatic vessels, and so back into the blood again, at the rate of a litre and a half a day.

This chapter elaborates on this succinct description.

Extravascular circulation of fluid and protein

The balance of forces governing fluid movement across the capillary wall, popularly known as the 'Starling principle' involves hydrostatic and colloid osmotic pressures both inside and outside the blood capillary. The principles of this exchange (Starling, 1896, 1898) rely on the relative impermeability of the capillary wall to protein molecules. A slow transcapillary movement of plasma protein from blood to interstitial space is constantly occurring (Landis & Pappenheimer, 1963). There are wide regional variations in the permeability of micro-vessels to plasma proteins and the formation of tissue fluid and lymph is predictable from the Starling principle applied to the individual tissue. Thus a vessel which is highly permeable to proteins, such as the hepatic sinusoid, will only achieve some degree of balance if its intravascular hydrostatic pressure is very low (see Chapter 8). Capillaries with low hydrostatic pressure and relatively low permeability to proteins, such as the pulmonary microvasculature, will tend to attract fluid into the intravascular compartment and conversely the high hydrostatic pressure of renal glomerular capillaries will overcome the plasma colloid osmotic pressure and result in a large filtration of fluid into Bowman's capsule.

A glance at Table 3.1, which shows the estimated proportion of the intravascular plasma protein which passes through the thoracic duct per day, will underline the importance of the dictum of Drinker (1946) that the most important general function of the lymphatic system is the 'unremitting removal of plasma proteins from the tissue spaces'. This clears the interstitial space of protein which otherwise would increase in concentration to the point where no gradient of colloid osmotic pressure would exist across the capillary wall and fluid loss driven by hydrostatic pressure from the circulation to the tissues would be unopposed.

TABLE 3.1. *Estimated percentage of total plasma protein passing through the thoracic duct per day in different species*

Species	% total plasma protein	Reference
Rat	66	Forker, Chaikoff & Reinhardt (1952)
Rabbit	62	Courtice & Morris (1955)
Cat	84	Courtice & Morris (1955)
Dog (anaesthetised)	55	Forker *et al.* (1952)
Cow	100	Hartmann & Lascelles (1966)
Calf	200	Shannon & Lascelles (1968a)
Man	50	Yoffey & Courtice (1970)

The formation of lymph

General considerations

Lymph is a component of the extracellular fluid of the body and is in large part derived from fluid and solute filtered from the blood circulation across the capillary wall. The volume and solute concentration of this filtrate is modified by passage through the tissues as interstitial fluid and the nature of the final product, lymph, is also determined by active participation of the endothelium of initial lymph vessels in the uptake of solutes and particles into the lumen of these vessels. The closer a lymph vessel is to a capillary blood vessel the more likely is the lymph in that vessel to approximate to the capillary filtrate (Clark & Clark, 1937) and despite the other factors governing lymph formation, regional lymph composition is frequently taken to reflect the composition of the capillary filtrate in the area drained. The concept of a balance of hydrostatic and osmotic forces governing fluid movement across the capillary wall described by E. H. Starling (1896, 1898) has been elegantly substantiated by the micro-injection experiments of Landis (1927a, b; 1930) studying frog and mammalian mesenteric capillaries and by Pappenheimer & Soto-Rivera (1948) using the isogravimetric preparation. In the perfused hind-leg of the cat the mean pressure governing the transfer of fluid across the capillary wall is the mean capillary hydrostatic pressure less the effective colloid osmotic pressure of the plasma proteins (Fig. 3.1). These two primary forces are quantitatively more important than their extravascular counterparts, the colloid osmotic pressure of interstitial fluid and the hydrostatic pressure of this fluid. Nevertheless these tissue fluid forces cannot be neglected and must be subtracted from the two major forces to determine the net forces operating across the capillary wall. As already mentioned the relative importance of these two tissue fluid pressures becomes greater in certain organs due to regional peculiarities of the microcirculation. For example, in the liver, sinusoidal pressure is low, about 4–7 mmHg, but the wall of the sinusoid is highly permeable to protein, thus the effective colloid osmotic pressure of the plasma is correspondingly reduced. In this situation a slight rise in sinusoidal hydrostatic

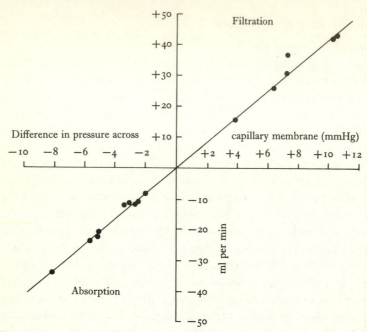

Fig. 3.1. Fluid exchange in the perfused hind-leg of the cat. The rate of fluid exchange is proportional to the difference between the mean hydrostatic pressure in the capillaries and the sum of all pressures opposing filtration and is independent of the absolute values of these quantities. Negative fluid exchange shows fluid entering the capillary, positive fluid exchange shows fluid leaving the capillary. (Redrawn from Pappenheimer & Soto-Rivera, 1948.)

pressure will cause a large filtrate to pass into the perisinusoidal space of Disse (see Chapter 8). It is important to consider that the hydrostatic pressure within a blood capillary is by no means constant and is regulated by arteriolar tone and the state of the pre-capillary and post-capillary vessels which govern the flow and pressure in the vessel. For a full discussion of general principles of lymph formation the reader is referred to the text of Yoffey & Courtice (1970).

Ultrafiltration and diffusion across capillary walls
The volume flow of solution across the capillary wall by ultra-filtration is given by:

$$Q_V = (P - P^1) - (P_o - P_o^1) \, L_p A,$$

34

where Q_V is the volume flow of solution across the capillary wall (filtration flow); P is the hydrostatic pressure in the capillary; P^1 is the hydrostatic pressure of the tissue fluid; P_o and P_o^1 are the effective osmotic pressures exerted inside and outside the capillary; L_p is the filtration coefficient and A the area of capillary membrane.

The filtration coefficient of a porous membrane such as the capillary wall can be represented as

$$L_p = \frac{N\pi R_p^4}{t_p 8\mu},$$

where N is the number of pores per unit membrane; μ is the viscosity of water; t_p is the pore length and R_p is the pore radius.

The above equations for the flow of ultrafiltrate only allow an approximate measurement as they apply to the idealised situation where the capillary wall is impermeable to protein.

Materials also cross capillary walls by diffusion, a very rapid process for small molecules such as water, ions, urea, glucose and gases. The rate of this process is given by:

$$M = D \frac{A}{\Delta_x} (C_1 - C_2),$$

where M is the net quantity of solute diffusing from the capillary per unit time, C_1, C_2 are the solute concentrations inside and outside the capillary, D is the free diffusion coefficient of the solute and A/Δ_x is the effective diffusion area per unit path length in the capillary wall.

$D\,A/\Delta_x$, therefore, represents the capillary permeability to the solute. A/Δ_x, is inversely related to the molecular size of the solute and becomes exceedingly small close to the molecular size of albumin whose effective diffusion radius is 3.5 nm. Small lipid-soluble molecules diffuse through capillary walls much more rapidly than water-soluble substances.

Study of capillary permeability to a range of molecules of increasing size shows a steeply graded reduction in permeability with increasing molecular size approaching zero at a molecular radius equal to that of a capillary pore radius of approximately 4 nm (Landis & Pappenheimer, 1963). Larger molecules filtered from plasma do appear in lymph and to account for this it is postulated that these molecules leak through a few large pores of about 80 nm radius or are transported in pinocytotic vesicles of radius approximately 25 nm (Garlick & Renkin, 1970). There is general agreement about the size of the small pores but various authors give widely differing estimates for the dimensions of the large pores or 'leaks'.

Furthermore, the nature of these large 'leaks' is far from clear. On the basis that the dye T-1824 is bound in the circulation to albumin, studies by Landis (1964) suggested that large 'leaks' exist in venous capillaries and venules in perfused frog mesenteric capillaries. Re-evaluation of this phenomenon by Levick & Michel (1973) showed that while such leaks can be demonstrated under conditions similar to those used by Landis, they probably represent the escape of unbound T-1824 which was present in substantial amounts in Landis' studies. The free molecule behaves as a small molecule with a Stokes–Einstein radius, calculated from the free diffusion coefficient, of 1.31 nm compared with 3.62 nm for the T-1824–albumin complex. Thus Landis' observations can be interpreted as evidence for non-uniformity of permeability of the capillary wall to small molecules but cannot be taken as evidence for large leaks.

The well-known differences in capillary permeability to various macromolecules is likely to represent a difference in the ratio of small pores to large 'leaks', whatever these 'leaks' may be. It is at the blood capillary wall that the principal barrier to the transport of large molecules from plasma to lymph exists (Garlick & Renkin, 1970).

Interstitial fluid pressure and lymph flow
Recent studies indicate that interstitial fluid is not a free fluid medium percolating slowly through the tissues. Rather, the interstitial compartment appears to consist of a mucopolysaccharide gel in which collagen and elastin fibres are embedded. This gel is unsaturated and tends to imbibe water (Laurent, 1970). The hydrostatic pressure of interstitial fluid plays an important part in determining lymph flow since it is a component of the Starling equation and therefore partly

determines the rate of fluid movement from blood to the extravascular space. Lymph is formed from interstitial fluid by forces which are still a matter of speculation.

Much controversy surrounds the assessment of interstitial fluid pressure. Several methods have been employed to measure this pressure. These include techniques using hypodermic needles (see, for example, McMaster, 1946). Pressure measurements of fluid accumulating in perforated plastic capsules implanted in subcutaneous tissues (Guyton, 1963) suggest that interstitial fluid pressure is negative with respect to the atmosphere. A mean value from several subcutaneous sites of −6.4 mmHg was obtained. Scholander, Hargens & Miller (1968) have also obtained negative values for interstitial fluid pressure with the use of impregnated cotton wool wicks. All these methods leave doubt about the significance of the results obtained since each of these techniques grossly interferes with a delicate system. Guyton *et al.* (1971) criticise the use of hypodermic needles on the grounds that a relatively large bore needle is used and also that injection of even a small amount of fluid causes distortion of the tissues. It is suggested that this technique merely estimates elastic rebound of the tissues. The capsule and wick techniques introduce a foreign body into the tissues. The capsule provokes the formation of new tissue around itself; thus a highly artificial situation is created and it is hazardous to assume that either pressure or composition of the fluid accumulating within the capsule is representative of the condition of normal interstitial fluid.

If interstitial fluid pressure is indeed negative with respect to the atmosphere, then it is necessary to explain how such a negative pressure is generated. No satisfactory explanation is available. It is proposed by Guyton *et al.* (1971) that the negative pressure is achieved by a combination of factors which include a 'sucking' action at initial lymphatic endothelial cell junctions and mechanical forces in the tissue generated by movement and vascular pulsation. In fact they have postulated that it is only necessary for lymphatics to clear the interstitium of protein rendering the interstitial fluid hypotonic, since the colloid osmotic pressure of the plasma will add to the negative pressure in the interstitial space. The 'sucking' action of initial lymphatics, however, which is central to this explanation is not clearly demonstrated or understood. A study by Taylor, Gibson & Gaar (1970) showed that as tissue fluid pressure rises from −6 mmHg to zero, as measured by the capsule technique, lymph flow is greatly increased. However, above this, further increases in lymph flow do not occur. Guyton and his colleagues (1971) explain this as due to the fact that as pressure rises from very negative values the lymphatic pump becomes more efficient until tissue pressure on the wall of the lymphatic vessel impedes further transport of fluid and a balance is struck between these two forces.

Another difficult problem is raised by the demonstration of slightly positive pressures in initial lymphatics by Zweifach & Prather (1975) using micropipettes. Interstitial fluid would have to overcome a considerable hydrostatic pressure gradient to enter an initial lymph vessel unless either a substantial colloid osmotic pressure gradient exists across the lymphatic wall due to higher protein concentration in lymph in that vessel than in the interstitial fluid or else transient reversals of pressures in tissue and lymphatics favours intermittent entry of fluid into the vessel. These possibilities are discussed in the section on the propulsion of lymph.

Formation of intestinal lymph
Though considerably lower than in hepatic lymph, the concentration of protein in intestinal lymph is relatively high (Table 3.2) implying that interstitial fluid protein concentration in the intestine is also quite high, presumably due to a moderate permeability of the intestinal capillaries to plasma protein.

Johnson & Hanson (1962) using indirect isogravimetric methods have found that the capillary hydrostatic pressure in the intestine is about 7 mmHg lower than the plasma colloid osmotic pressure and postulated that this imbalance is offset by a substantial tissue oncotic pressure. This notion is supported by the conclusion of others that interstitial fluid protein concentration is on the average about 2.1 g per 100 ml (Landis & Pappenheimer, 1963). Thus the oncotic pressure of tissue fluid is about 5 mmHg on average and may well be considerably

TABLE 3.2. *Protein concentration of dog lymph from various regions*

	Total protein		Albumin		Globulin		Reference
	Plasma	Lymph	Plasma	Lymph	Plasma	Lymph	
Thoracic duct	6.19	4.00	3.56	2.45	2.62	1.54	Field et al. (1934–5)
	5.91	3.23	3.33	2.04	2.08	0.88	Nix, Flock & Bollman (1951)
	5.65	3.44	3.67	2.38	1.97	1.08	Courtice & Morris (1955)
Liver	6.34	5.32	3.38	2.81	2.96	2.51	Field et al. (1934–5)
	6.67	4.39	3.41	2.74	1.81	1.28	Nix et al. (1951)
Intestine	6.24	3.98	3.67	2.42	2.57	1.56	Field et al. (1934–5)
	5.98	2.97	3.18	1.72	2.80	1.25	Wells (1932)
	5.67	2.79	3.47	1.90	1.62	0.64	Nix et al. (1951)
Cervical duct	6.25	3.63	3.61	2.36	2.63	1.26	Field et al. (1934–5)
	5.65	2.57	3.67	1.72	1.97	0.85	Courtice & Morris (1955)
Leg	6.46	1.91	3.62	1.20	2.84	0.71	Field et al. (1934–5)

greater in the intestine. In the cat mesentery, direct micro-puncture experiments have shown a mean capillary hydrostatic pressure of 24 mmHg and a blood oncotic pressure of 20 mmHg. Peripheral lymph hydrostatic pressure was estimated as zero with a lymph oncotic pressure of 6 mmHg (Hargens & Zweifach, 1976). Alterations in the blood variables led to rapid compensatory changes in tissue variables thus maintaining the Starling balance (see Table 3.3).

The intestine has a great capacity to neutralise large changes in capillary pressure as judged by isogravimetric (Johnson & Hanson, 1963) and isovolumetric studies (Wallentin, 1966). Despite substantial increases in venous outflow pressure the intestine seems to be protected against oedema in these preparations. In these studies lymphatics were ligated so the observations imply that substantial adjustments in some component or components of the Starling equilibrium have compensated for the increased capillary hydrostatic pressure. Wallentin (1966) has proposed that a rise in tissue fluid hydrostatic pressure opposes outward flow while Johnson & Richardson (1974) believe that adjustments in tissue fluid oncotic pressure play a primary role in limiting blood–tissue fluid exchange. In dog intestinal lymph they found that protein concentrations fell from 5.1 % ±0.8 % when venous pressure was 0 mmHg to 0.8 % ±0.3 % when venous pressure was raised to 25 mmHg. These observations, together with the fact that at the higher pressures the weight of the preparation started to rise as capillary pressure exceeded the net colloid osmotic pressure, support the view that changes in tissue fluid oncotic pressure play an important part in limiting fluid transfer across the intestinal capillaries, but it is debatable whether one can conclude that tissue fluid oncotic pressure is cause rather than effect in such situations. Nevertheless these preparations reveal underlying autoregulatory mechanisms which are modified by other factors in the intact animal which include adjustments in arteriolar tone in response to increase in venous pressure – the 'venous-arteriolar' response (Johnson, 1959, 1960). This is a myogenic response of pre-capillary resistance vessels which constrict in response to stretch caused by increased pressure.

The intestine, like other tissues, is further protected from

TABLE 3.3. *Effect of diluting blood 1:1 with buffered saline (perturbation) on 'Starling factors' in cat mesentery.* (Modified from Hargens & Zweifach, 1976.)

Values represent means±s.D. for two cats, *n* = number of individual measurements.

	Blood capillary hydrostatic pressure (mmHg)	Blood colloid osmotic pressure (mmHg)	Lymph colloid osmotic pressure (mmHg)	Lymph flow (ml per min per 100 g)	Capillary filtration coefficient (ml per min per 100 g per mmHg)	Net filtration pressure (mmHg)
Normal	22±1 (n = 8)	19.1±0.5 (n = 4)	4.5±0.2 (n = 4)	0.018±0.004 (n = 10)	0.0024	7.4
Post-perturbation 1 hour	22±1 (n = 6)	8.8±2.4 (n = 2)	1±0.3 (n = 2)	0.170±0.010 (n = 18)	0.0121	14

accumulation of interstitial fluid by the flexibility of lymph flow (Hargens & Zweifach, 1976). Lymph flow contributes to a 'safety factor' which combats oedema. If interstitial fluid pressure exceeds 3 mmHg, fluid moves into the intestinal lumen (Wilson, 1956). On the basis that intestinal lymph flow can increase by at least tenfold (Korner, Morris & Courtice, 1954) and can reduce the interstitial space protein concentration by 50%, Taylor, Gibson, Granger & Guyton, (1973) have estimated that a total safety factor of at least 20 mmHg must be overwhelmed before the mucosa secretes fluid into the intestinal lumen. Further data on this subject have been produced by Yablonski & Lifson (1976) who have shown in dogs that small intestinal fluid secretion begins when venous pressures exceed 22–26 mmHg and that above this threshold secretion increases in proportion to pressure. Lymph flows also increase at roughly the same rate as this secretion (Fig. 3.2) and in agreement with the observations of Wilson (1956) it seems that only when mucosal pressures exceed 3–4.5 mmHg

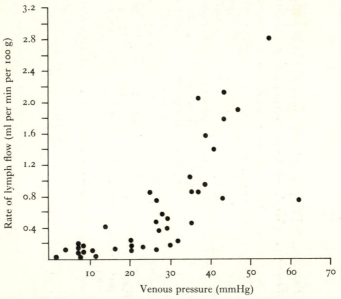

Fig. 3.2. Relationship between the rate of lymph flow and intestinal venous pressure in a loop of dog small intestine *in vivo*. (From Yablonski & Lifson, 1976.)

is secretion initiated, driven by this head of pressure and facilitated by an increase in hydraulic permeability of the epithelial layer of the mucosa.

When followed over some hours, however, the relationship between intestinal lymph flow and intestinal secretion appears to be rather complex. Using an isolated vascularly perfused cat ileal preparation, Granger, Mortillaro & Taylor (1977) showed that perturbation of the Starling factors either by raising venous pressure or reducing plasma oncotic pressure leads to large but transient rises in intestinal lymph flow. In response to raising venous pressure to 30 mmHg lymph flow rose by 20 times control values and infusion of Tyrode solution (2.5 ml per min per kg) into the femoral vein increased lymph flow to 38 times control. These rises, however, were transient and flow rates returned to basal levels over two to three hours. As the flow rate fell intestinal secretion rate rose. An explanation for these effects might be that at a certain pressure in the mucosal interstitium a low-resistance fluid pathway is opened, possibly by disruption of the tight junctions between mucosal epithelial cells, thus decompressing the interstitium and reducing lymph flow.

The 'safety factor' of the intestine is clearly considerable and is important since this tissue is subject to large increases in blood flow and capillary filtration during periods of activity and also subject to large influxes of fluid and solute from the gut lumen. The small intestine has a considerable capacity to remove albumin from the interstitial space, probably as a result of its high lymph flow. Sheppard & Sterns (1975) have shown that when subserosal injections of radio-iodinated serum albumin are given to dogs with thoracic duct fistulae the stomach and colon clear about 9% by the lymphatic route in five hours but the small intestine clears about 37% by this route during the same period.

When portal venous pressure rises intestinal lymph flow increases and transport of fluid and solute by this route becomes more important. Morris (1956b) found in the resting cat intestine a lymph flow of about 4 ml per hour. When the portal venous pressure was increased by 12 cm of water (9 mmHg) the lymph flow rose, slowly, to 6 ml per hour. The studies of Lee &

Duncan (1968; see Chapter 6) emphasise the importance of intestinal lymph in clearing absorbed fluid as portal venous pressure is raised. The interstitial fluid of the intestine probably acts as a common pool in which fluid and solute absorbed from the intestine and plasma filtrate mix and from which the individual components are taken up by the fast-flowing portal blood and the slow-flowing lymph. The factors governing the route taken by any of these components include molecular or particle size and hydrostatic and osmotic pressure gradients between interstitium and vessel lumen. This concept is discussed further in Chapter 6.

Intestinal blood flow and lymph formation
Intestinal mucosal circulation. Though difficult to prove, it is likely that a major contribution to intestinal lymph comes from the capillary filtrate in the lamina propria of the small intestinal mucosal villi. The anatomical arrangement and ultrastructural characteristics of this circulation are, therefore, of great interest.

The vascular architecture of villi varies greatly among different species (Sessions, Viegas de Andrade & Kokas, 1968). The variations depend mainly on whether the arteriole supplying the villus passes unbranched up the villus in the axial or para-axial position to form a cascade of capillaries running subepithelially to a draining venule at the base or whether the arterial vessel arborises at the base of the villus and the capillaries pass up the villus to collect in venules near the tip. Clearly, the direction of capillary flow in the former group is central and in the latter, distal. In either case, the capillary flow is counter to the main supplying (arteriole) or draining (venule) vessel. Conditions are thus satisfied for a possible intestinal mucosal counter-current exchanger (see below). In some species such as the cat (Casley-Smith, O'Donoghue & Crocker, 1975), an intermediate arrangement of the microvascular anatomy of the villus exists. Thus, the arterial vessel divides into capillaries at the base of the villus and the villous capillaries ascend to the tip to form larger draining 'venous' capillaries which later merge as a post-capillary venule about halfway down the villus.

44

Ultrastructure and permeability of intestinal capillaries. Like the capillary vessels of many specialised tissues including endo-crine and exocrine glands and the renal glomerulus, intestinal capillaries are fenestrated (Plate 3.1). The fenestrae are circular having a diameter of 35–45 nm. The diaphragm of the fenestra is in continuity with the outer leaflet of the plasmalemma and often has a central knob of about 15 nm (Clementi & Palade, 1969). In both mouse and cat intestinal capillaries, fenestrae are much more frequent at the venous end than the arterial side (Casley-Smith, 1971; Casley-Smith *et al.*, 1975). This is the area which shows the greater permeability to water-soluble dyes (Rous, Gilding & Smith, 1930), albumin (Landis, 1964) and water (Wiederhielm, 1967; Intaglietta, 1967). In the mouse, Casley-Smith (1971) has observed that the fenestrae of sub-epithelial capillaries are chiefly on the epithelial side of the vessel and the nucleus is on the opposite side.

There is some evidence which suggests that fenestrae on the venous side of capillaries may be a route for return of extravasated albumin to the blood (see Casley-Smith, 1972; 1976); it is suggested that proteins leave the circulation through fenestrae at the arterial end of the capillaries and thus a local pericapillary extravascular circulation of plasma proteins would occur. This raises the question of the magnitude of the return of plasma protein to the circulation by direct uptake into blood capillaries. Perry & Garlick (1975) have demonstrated direct uptake of iodinated gamma globulin from the interstitial space of an isolated perfused rabbit gastrocnemius muscle into blood. This preparation employs an artificial plasma consisting of albumin and dextran and though labelled protein enters the blood vessels from the interstitium of the muscle, it is difficult to relate these findings quantitatively to the intact system. Nevertheless, the fact that proteins absorbed from the intestine or synthesised in intestinal mucosa are taken up by both blood and lymph (see chapter 6) indicates that both these routes are involved in the transport of protein from interstitial fluid.

The quantitative importance of backflux of proteins from the interstitial space to blood capillaries in terms of the whole body has been questioned by Lassen, Parving & Rossing (1974).

45

From calculations of trans-capillary escape of albumin by measurements of escape of labelled albumin from plasma and albumin transport in the thoracic duct, it would appear that outflux of albumin is only about 1.2 times the total lymphatic return of albumin. This suggests that in whole body terms, the return of plasma protein across the capillary wall is quantitatively slight and that proteins which have once left the circulation by filtration are destined to return to the blood via lymph. The extent to which tissue cells take up and metabolise interstitial fluid proteins is an important consideration though it is difficult to assess its magnitude. The forces responsible for the uptake of protein from the interstitial fluid into the blood remain to be defined.

Clementi & Palade (1969) have attempted to identify structural equivalents of the large and small pores held to account for permeability characteristics of the intestinal microcirculation (see above). Horseradish peroxidase (molecular diameter ~ 5 nm) and ferritin (molecular diameter ~ 11 nm) were used as probe molecules. Between 1 and 1½ min after intravenous injection of peroxidase, collections of the enzyme could be identified in the pericapillary spaces, the highest concentration being opposite fenestrae. Similar results were found with ferritin. Neither tracer appeared to leave the bloodstream through intercellular junctions, and vesicular transport, though present, was quantitatively much less important than passage through the fenestrae. The molecular dimensions of the probe molecules might have been expected to distinguish small and large pores as 9 nm is taken as the upper limit for the diameter of the small pores (Grotte, 1956). The findings in Clementi & Palade's study do not distinguish morphologically the two pore systems. It is possible that small pores exist in the diaphragms of fenestrae and that occasional fully opened fenestrae, whose diameters are approximately 35–45 nm account for the escape of the larger molecule. In non-fenestrated or continuous capillaries the small pores of Pappenheimer (1953) probably correspond to 'close' endothelial cell junctions, the gap being about 6 nm (Casley-Smith, 1976). Proteins probably leave these capillaries by transendothelial vesicular transport.

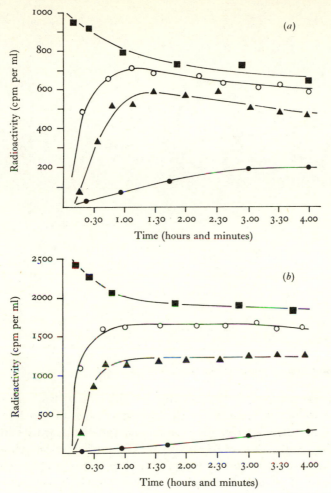

Fig. 3.3. The disappearance of dextran of MW 35000 (*a*) and ¹³¹I-labelled albumin (*b*) from plasma and its appearance in lymph in the dog. ■———■, plasma; O———O, hepatic lymph; ▲———▲, intestinal lymph; ●———●, cervical lymph. (From Mayerson *et al.*, 1962.)

The relatively high concentration of protein in intestinal lymph suggests that the intestinal capillary bed is fairly permeable to macromolecules, and studies in dogs by Mayerson, Wolfram, Shirley & Wasserman (1960); and Mayerson *et al.* (1962) show that the intestinal capillary bed is less permeable than the hepatic sinusoids to macromolecules but much more permeable

47

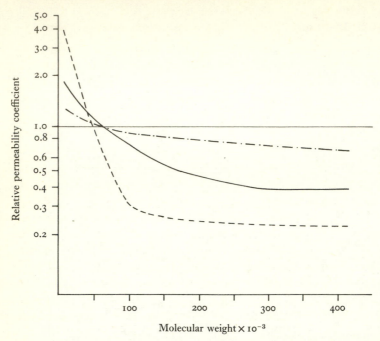

Fig. 3.4. The permeability coefficients, relative to albumin, of dextrans of different molecular weight for hepatic, - — · — ; intestinal, ——— and cervical, - - - - - - tissues. Mean values for groups of dogs. (From Mayerson *et al.*, 1962.)

to macromolecules than the capillary bed drained by the cervical lymph duct, as judged by the disappearance of labelled albumin and dextran from the plasma and their appearance in lymph (Fig. 3.3). This is further substantiated by comparing the ratio of concentration in lymph : concentration in plasma for dextrans of graded molecular weight relative to the same ratio for albumin in various vascular beds (Fig. 3.4).

Reference has already been made to the pore theory of capillary permeability. Grotte (1956) using dextrans has calculated a ratio of 1:340 for large pores to small pores in the hepatic sinusoids. Estimates for the ratio in limb capillaries are around 1:30000 and the ratio for intestinal capillaries can be presumed to be intermediate though closer to the value for the hepatic sinusoid.

Intestinal blood flow and gastro-intestinal function

Many studies indicate that superior mesenteric blood flow increases and mesenteric vascular resistance decreases after a meal (Herrick, Essex, Mann & Baldes, 1934; Brandt *et al.*, 1955; Reininger & Sapirstein, 1957; Fronek & Stahlgren, 1968; Vatner, Franklin & Van Citters, 1970). Assuming that this implies increased flow through capillary beds in the gut it could be expected that lymph production in these tissues will rise. The increase in blood flow begins five to fifteen min after eating, reaches a maximum in about one hour usually about 100–200% above control values and returns to basal levels in three to seven hours (Vatner *et al.*, 1970). The mechanisms involved in these changes are not clear.

The intestinal circulation is under both neural and humoral control. Quantitative studies on a denervated intestinal segment in the anaesthetised fasting cat show that total intestinal blood flow is 20–40 ml per min per 100 g intestine (Lundgren, 1967) and villus blood flow is approximately 30–60 ml per min per 100 g of villus tissue (Jodal & Lundgren, 1970). Biber, Lundgren & Svanvik (1973) have estimated that approximately 80% of total intestinal blood flow passes to the mucosa and submucosa and of this about one-third passes to the villi.

Although cholinergic fibres are found in association with mesenteric vessels the major autonomic innervation of these vessels is a dense supply of adrenergic fibres. Their stimulation can reduce splanchnic blood flow to zero though escape occurs with continued stimulation and blood flow returns to values a little below basal levels (Folkow *et al.*, 1964). This auto-regulatory escape seems to involve a redistribution of intestinal intramural blood flow with villus blood flow maintained constant or slightly increased (Svanvik, 1973). It seems that humoral agents play a more important role than autonomic nerves in the mesenteric vasodilation which accompanies digestive activity. These agents probably include histamine, 5-hydroxytryptamine and several gastro-intestinal hormones. Secretin (Ross, 1970*a*), cholecystokinin (Fara, Rubinstein & Sonnenschein, 1969; Dorigotti & Glässer, 1968) glucagon (Ross, 1970*b*) and vaso-active intestinal peptide (Said & Mutt,

49

1970) all have potent dilator action on the mesenteric circulation. Both secretin and cholecystokinin increase intestinal capillary filtration (Biber, Fara & Lundgren, 1973*a*) though secretin appears to increase submucosal blood flow while cholecystokinin enchances mucosal blood flow (Fara & Madden, 1975). In anaesthetised dogs, secretin and cholecystokinin have been shown to enhance thoracic duct flow (Razin, Feldman & Dreiling, 1962).

The mesenteric vasodilatation which follows a meal is blocked by atropine but not by bilateral thoracic vagotomy or by α- or β-adrenergic blockade (Vatner *et al.*, 1970). This is compatible with the hypothesis that feeding releases humoral vasodilator agents by local cholinergic mechanisms. Fara, Rubinstein & Sonnenschein (1969) showed that milk introduced into the duodenum in cats leads to a 50 to 100% increase in superior mesenteric blood flow in three to six minutes. Instillation of fat, L-phenylalanine or acid into cat duodenum leads to a significant rise in gall bladder pressure and pancreatic enzyme secretion together with an increase in mesenteric blood flow. These changes could be explained on the basis of release of endogenous cholecystokinin and secretin. Cross perfusion experiments by Fara *et al.* (1972) lend further support to this humoral hypothesis since introduction of lipid into the duodenum of a donor cat produces enhanced superior mesenteric flow in both donor and recipient. The vasodilatation after infusion of fat, amino acids or hydrochloric acid into the duodenum is mainly confined to the superior mesenteric bed and this selectivity of effect has been ascribed by Fara and his colleagues to secondary vascular changes consequent on a metabolic effect of the hormones on their target tissues, i.e. increased intestinal motility, increased tone in gall bladder smooth muscle and increased metabolic activity in the exocrine pancreas associated with enzyme secretion. This conclusion is supported by finding a correlation between jejunal and pancreatic blood flows and the oxygen consumption of these tissues.

The mechanisms of mesenteric vasodilatation after feeding probably involve several humoral agents acting directly on blood vessels or indirectly through the gastro-intestinal tissues. Local mechanisms initiated by mechanical and chemical stimuli to the intestinal mucosa are also probably involved (Biber, Lundgren & Svanik, 1973). The marked increase in intestinal lymph flow and plasma protein transport in lymph during digestion and absorption of fat meals (see Chapter 7) reflects such changes in mesenteric blood flow.

Distribution of blood flow in the intestine

Two models are proposed for the vascular arrangements in the intestine. In one, the blood supply to the mucosa and submucosa is linked in series (Ross, 1971; Greenway & Murthy, 1972) and in the other, muscle, submucosal and mucosal blood vessels are lined in three parallel-coupled circuits (Folkow, 1967; Lundgren, 1967). A number of studies suggest that considerable redistribution of blood within the layers of the intestinal wall can occur under different circumstances. Using ^{85}Kr washout measurements, Lundgren (1967) has found that vasodilatation induced by isopropylnoradrenaline produces marked increases in flow to the submucosal region. Fara & Madden (1975) using radioactive microspheres of 15 μm diameter have shown that secretin enhances intestinal blood flow predominantly by enhancing submucosal flow while cholecystokinin increase flow predominantly to the mucosa. As both hormones increase intestinal capillary filtration, these observations would support the concept that filtration can occur in small vessels in the submucosa and this area will become an important source of intestinal lymph when the blood is shunted there.

Counter-current exchange and intestinal blood flow. In 1967 Lundgren studying the washout from the small intestine of ^{85}Kr administered intra-arterially observed a large rapid component of the washout suggestive of shunting of the tracer within some sections of the small intestinal wall (Kampp, Lundgren & Nilsson, 1967). However Grim & Lindseth (1958) using radioactively labelled glass spheres had already shown that arteriovenous anastomotic flow in the gut only amounts to 3–4% of the total intestinal flow. An alternative explanation for Lundgren's observations is the existence of a counter-current exchange mechanism operating in the mucosa of the small intestine.

The anatomical arrangement of the blood vessels of the intestinal villus meets the structural prerequisites for a counter-current exchanger in so far as the arterial and venous limbs of the vascular loop are in close proximity, approximately 15–20 μm as estimated by Lundgren (1967). Measurements of the 'shunting' effect in ^{85}Kr washout curves at different intestinal blood flows produced by vasodilatation with infusions of iso-propylnoradrenaline showed that the fast component of the washout seemed to move from the region of the crypts to the tips of the villi as blood flow increased from 20 to 100 ml per min per 100 g. At low rates of blood flow intravascularly administered lipid-soluble drugs such as antipyrine were excluded from the villi. A similar but less pronounced effect was seen with lipid-insoluble substances. These observations are in keeping with the operation of a counter-current exchange mechanism in the villi and the faster the linear blood flow along the hairpin loop the smaller becomes the counter-current exchange of solutes. At blood flows in excess of about 150 ml per min per 100 g the counter-current exchanger probably does not operate (Lundgren, 1967).

An important functional implication of this is that substances transferred by either passive or active absorption across the intestinal epithelium into the villus tip and venous vessels will participate in this exchange and diffuse into the central ascending arterial vessel thus delaying absorption (Fig. 3.5a). The effectiveness of the system in delaying absorption will depend on the proximity of the limbs of the hairpin loops, the permeability of their walls, the linear velocity of blood flow in the loop and the diffusion characteristics of the solute.

Blood-borne substances which have a higher arterial than venous concentration on arrival at the intestinal mucosal circulation may exchange as shown in Fig. 3.5b. Thus this exchanger, by reducing the interstitial concentration of the substance progressively from crypt to villus tip, might reduce the loss of such substances to the intestinal lumen. Oxygen is one such substance which is shunted at low rates of blood flow (Kampp, Lundgren & Nilsson, 1967); thus there is a graded decrease in oxygen pressure in the villi from crypts to their tips. It is likely that lymphatic concentrations of substances

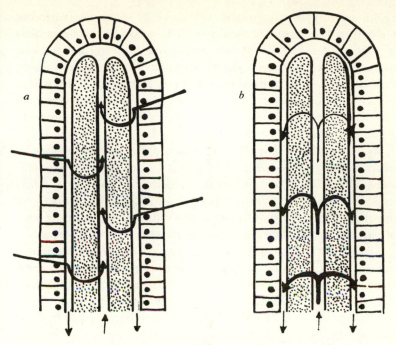

Fig. 3.5. Functional implications of the mucosal counter-current exchanger schematically illustrated. The intervascular distance is greatly exaggerated for the sake of clarity. (*a*) Shows the behaviour of a solute absorbed either actively or passively from the intestinal lumen into venous vessels of the villus and thence into the arterial vessel; (*b*) illustrates the behaviour of a solute delivered to the villous circulation via the arterial blood. (From Lundgren, 1967.)

exchanged in this way will be in equilibrium with intravascular and interstitial concentrations.

Measurements of sodium concentration in the villi of cat small intestine at 'resting' intestinal blood flow during the absorption of sodium chloride from hypo- or hypertonic solutions show that the sodium concentration per milligram tissue protein is three to four times greater at the villus tip than at its base (Haljamäe, Jodal & Lundgren, 1971; Haljamäe, Jodal, Lundgren & Svanik, 1973). This gradient is abolished in the presence of intense vasodilatation and this is in keeping with a counter-current multiplication of sodium occurring between the limbs of the hairpin loop. This could create the osmotic force enabling water transport to occur in the absence of, or against, an *apparent* osmotic difference. The concentration of sodium in the central lacteal would be expected to be similar to that in the interstitial space but the hypertonicity of the tissue fluid at the villus tip leads secondarily to increased tissue fluid pressure driving fluid into the central lacteal. The gradient of solute concentration in the interstitial fluid of the villus from tip to base would also create a similar hydrostatic pressure gradient promoting centripetal lymph flow.

Recently, Jodal (1977) has obtained further experimental support for this idea by demonstrating an osmotic gradient along the villus from tip to base: using cryoscopy to study intestinal sections a mean tissue osmolality of about 1100 mosmoles/kg water was found at the villus tip while at the base of the villus osmolality was comparable to that of plasma.

Based on the observations that a counter-current exchange in the mucosal circulation can trap solutes in the villus tip, Jodal (1973) has proposed that the fate of fatty acids absorbed from the gut is partly determined by this mechanism. It is suggested that long-chain fatty acids are not totally re-esterified on passage through the epithelial cell but that trapping in the villus circulation and interstitium allows time for these fatty acids to be incorporated into triglycerides which are finally absorbed into the lymphatic vessel. Water-soluble short-chain fatty acids, such as butyric acid, being pore-restricted in their passage through cells and capillary vessel walls, are much less affected by the counter-current exchange mechanism. In addition these fatty acids have much less affinity than long-chain fatty acids for re-esterifying enzymes in the mucosal cell (Brindley & Hübscher, 1966). Thus their absorption is relatively fast and their transport direct – via venous blood. This interesting hypothesis needs to be tested further.

The route of fluid, solutes and particles into lymphatics

The tenuous incomplete basement membrane of the lymphatic endothelium offers little barrier to the passage of solutes, fluids and large particles into fine lymph vessels. The incomplete nature of the basement membrane together with a relative paucity of adhesion devices between the endothelial cells compared with those of blood capillary endothelium allows occasional separation of those cells. The intercellular gaps so created would permit passage into lymphatics of a wide spectrum of material ranging down in size from whole cells.

There are three possible routes for entry of material into terminal lymphatics, i.e. between endothelial cells, in vesicles crossing the cell in which the substance would be separated from the cytoplasm by a bounding membrane, or by direct passage through the cytoplasm of the endothelium. Morphological techniques have been used in an attempt to demonstrate the route taken by a large variety of substances which have ranged in size from red blood cells down to various ions, and have included many types of particle of intermediate size such as carbon particles and thorium dioxide. Naturally occurring materials such as chylomicra and lipoproteins, which are ordinarily transported in quantity in lymph, have also been studied.

The appearance of initial lymph vessels is rather variable and appears to depend on the state of activity of local tissues. On a morphological basis two types of fine lymphatic capillary are described: one type which possesses few intercellular gaps is found in quiescent regions such as the pinnae of the ear and the other, with much more numerous intercellular gaps is found in active areas such as the diaphragm, intestinal villus and cardiac and active skeletal muscle. To some degree this may be a basic structural difference in these two types of vessel but it is also likely that local mechanical activity is responsible for drawing endothelial cells apart, creating the appearances seen in the second group. The importance of collagenous fibres attached to the abluminal surface of lymphatic endothelial cells in distracting these cells at their junctions has already been described (Chapter 2). The frequency with which open junctions in lymphatic vessels occur is the subject of some debate. It is likely that many such open junctions observed in histological preparations have arisen through unavoidable trauma to the tissue since it is well known that it is easy to demonstrate penetration of macromolecules between lymphatic endothelial cells following local trauma (Casley-Smith, 1964a).

Cells and large particles. With the light microscope Henry (1933) and Clark (1936) observed the passage of cells and large particles between lymphatic endothelial cells. Several studies have shown that chylomicra can use this route (see, for example, Casley-Smith, 1962; Sabesin, 1976; Plate 3.2). Chylomicra are also able to pass into and through endothelial cells in large vesicles (Casley-Smith, 1962; Dobbins, 1971).

Vesicular transport of foreign material is also known to occur in lymphatic endothelium, but vesicles involved are generally smaller than those enclosing chylomicra. Larger foreign particles of the size of chylomicra do not appear to be *transported* by vesicles but do seem to be taken up and retained in the cell. The transport of chylomicra across the cell in large vesicles therefore appears to be a specific phenomenon.

In an attempt to assess the relative importance of vesicular versus intercellular passage of particles and macromolecules into intestinal lymph vessels, Dobbins & Rollins (1970) used peroxidase, ferritin and chylomicra as markers and studied the frequency of the appearance of those markers in vesicles and in relation to intercellular junctions. They suggest that, with careful histological preparation, few open gaps are to be found between endothelial cells of lacteals in guinea pigs and mice. Adhesion devices between endothelial cells were frequently seen in their preparations and it was suggested that vesicular transport of these macromolecules and particles is quantitatively more important than the passage by the intercellular route. This view is challenged by Casley-Smith (1972) who, on the basis of a correlation of frequency of intercellular gaps and macromolecular permeability of different lymphatics, considers that the intercellular route is of paramount importance in the entry of particles, macromolecules and fluid into lymphatics. This is the popular view though the difficulties in quantifying the importance of the two routes should be remembered.

Small particles. This group of particles includes large molecules ranging in diameter between about 3 nm and 50 nm. It has been shown by electron microscopy that particles in this group can pass through intermediate junctions (zonulae adhaerentes) of the endothelial cells but tight junctions (zonulae occludentes) form a barrier to their passage. In situations where junctional complexes are absent, these particles will pass easily into the lumen of the vessel between the endothelial cells. Substances in this group can also be shown to be taken up by vesicles of approximately 50 nm diameter and morphological appearances suggest that these materials can be transported *across* the cell by this route. As has been already mentioned, larger vesicles containing foreign material not destined for lymphatic transport are seen in lymphatic endothelium. They are thought to form in two ways, either by coalescence of small vesicles containing the foreign substance or by uptake of a large amount of the material *en masse*.

Ions and small molecules. Using precipitation techniques, Casley-Smith (1967b) obtained evidence with the electron microscope that ions can enter the lymphatic lumen through zonulae occludentes. He calculated that the maximum size of openings through these tight junctions is approximately 3 nm. As well as this route through intercellular junctions, ions and small molecules are able to pass through the endothelium in vesicles along with larger particles and will also stream through any patent intercellular gaps present. It is difficult to give a quantitative assessment of the importance of these various routes for small molecular species. Small lipid-soluble molecules have an alternative route in so far as they can dissolve in the lipid membrane of the endothelial cell and be transferred to the lumen, either around the cell within the membrane, or through the membrane to the cytoplasm and across to the luminal aspect of the endothelial cell.

Fluid uptake. Water probably enters lymphatic vessels by all the routes described above. In a situation where water is accumulating rapidly in interstitial tissue, as for example when the intestinal mucosa is exposed to a hypotonic medium, the rise in interstitial pressure produced by water accumulation may cause distraction of lymphatic endothelium by stretching the connective tissue fibres linked to the abluminal surface of these cells. The poorly developed basement membrane of the

57

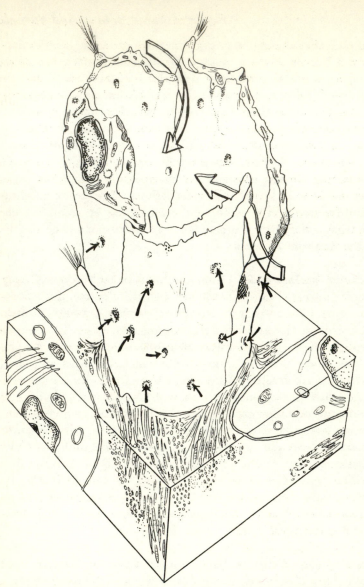

Fig. 3.6. Three-dimensional diagram of a short segment of a lymphatic capillary reconstructed from collated electron micrographs. The major passage way for the transport of fluids and particulate substances from the interstitium into the lymphatic lumen is the intercellular cleft (long white arrows). The uptake of particulate components from the connective tissue front is also carried out by vesicles containing the particles. It is suggested that the movement of particles proceeds from the connective tissue front of the endothelium via vesicles (small arrows). These vesicles seem to merge or fuse with autophagic vacuoles. (From Leak, 1976.)

58

lymphatic endothelium may allow some separation of the cells. Under these circumstances, bulk fluid flow into the lymphatic lumen would occur through intercellular gaps. Some water will be transported by vesicles, some will pass through intercellular junctional complexes, and probably some will pass directly through the cytoplasm of the endothelium. The routes of entry of fluid and solute into lymph vessels are illustrated diagrammatically in Fig. 3.6.

The composition of lymph

Proteins

All plasma proteins are found in lymph. Plate 3.3 and Fig. 3.7 compare the electrophoretic and molecular exclusion chromatographic patterns of rat plasma protein with intestinal lymph protein. The concentration of total protein in central lymph, that is the lymph from the cisterna chyli or thoracic duct, is the resultant of the protein concentrations of lymph from various regions, the highest concentration being in hepatic lymph (see Chapter 8) where concentrations approach those of plasma. Intestinal lymph, which also makes a large contribution to thoracic duct flow, has a somewhat lower protein content. There is a large amount of data on concentrations of protein in lymph from various regions and the reader is referred for details to the text by Yoffey & Courtice (1970, Chapter 4). The data for one species, the dog, have been collected in Table 3.2 (p. 39). The protein concentration of lymph from various parts of the alimentary tract tends to be quite high. For example canine gastric lymph, collected by micropuncture, contains all the plasma proteins as judged by polyacrylamide gel electrophoresis of lymph samples and plasma samples obtained at the same time (Bruggeman, 1975). On cellulose acetate electrophoresis quantitative analysis showed gastric lymph:plasma ratios of 0.51 for total protein, 0.68 for albumin, 0.42 for total globulin, 0.49 for α_1-globulin, 0.62 for α_2-globulin, 0.35 for β-globulin, 0.39 for fibrinogren and 0.46 for γ-globulin. If the concentration of albumin in gastric lymph reflects interstitial fluid concentrations of albumin in gastric mucosa, the high ratio for albumin suggests

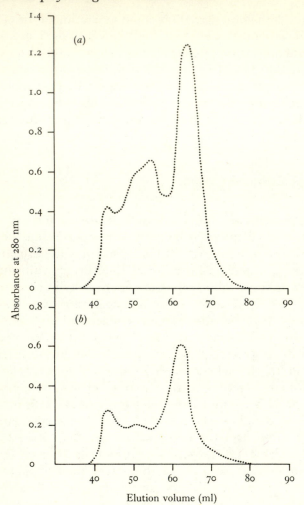

Fig. 3.7. Filtration of rat plasma and intestinal lymph on a column of Sephacryl S-200 (an allyl dextran covalently cross-linked with N,N-methylene bisacrylamide) eluted with o.1 M Tris-HCl buffer, pH 8.0 containing 1 M NaCl. (*a*) 0.5 ml plasma; (*b*) 1.0 ml intestinal lymph from which chylomicra had been removed by centrifugation. The macromolecular peak of lymph (elution volume 43 ml) is relatively large compared with the albumin peak (elution volume 62 ml). This may represent an enrichment of intestinal lymph with respect to immunoglobulins synthesised in the intestinal mucosa (see Chapter 6).

that there may be a comparatively small hydrostatic pressure gradient across gastric capillary blood vessels.

Information about the composition of human lymph comes from patients undergoing therapeutic drainage of the thoracic duct, or in earlier cases lymph obtained from subjects with accidental lymph fistulae (see, for example, Crandall, Barker & Graham, 1943; Courtice, Simmonds & Steinbeck, 1951; Bierman *et al.*, 1953; Linder & Blomstrand, 1958; Blomstrand, Dahlbäck & Radner, 1960; and Dumont & Mulholland, 1960). Bergström & Werner (1966) have made a systematic study of the chemistry of human lymph comparing it with that of blood. Thoracic duct lymph was found to have a protein concentration of about 66% that of serum. The ratios of the concentrations of albumin, total globulin and macroglobulin in lymph to those of blood serum were found to be about 0.75, 0.5 and less than 0.3 respectively while the fibrinogen content of lymph was approximately 25% that of blood plasma. Using thin-layer molecular exclusion chromatography, it was found that there was a good correlation between the molecular size of a protein and its lymph:serum ratio. Thus molecular size plays an important part in determining the quantities of protein crossing the blood:lymph barriers, reflected in higher albumin:globulin ratios in lymph than in plasma. A large number of specialised proteins have been detected in lymph. Enzymes present in plasma appear in lymph taking part in the extravascular circulation of protein (Yoffey & Courtice, 1970). Chemical concentrations of enzymes in blood plasma are low and lymph values tend to be rather lower though concentrations of enzymes derived from certain tissues such as the small intestine and pancreas may be relatively high in the lymph draining these organs (see Chapter 5). In human thoracic duct lymph, alkaline phosphatase, acid phosphatase, lactic dehydrogenase, glutamic–oxaloacetic transaminase, glutamic–pyruvic transaminase and aldolase have been measured and all except aldolase were in lower concentration in lymph than in blood. Lymph aldolase levels were approximately twice as high as blood levels (Werner, 1966*b*).

Specialised transport proteins of plasma also appear in lymph, and Morgan (1963) has shown that iron remains bound to transferrin during its escape from blood to lymph. The concentration of protein-bound iron in human thoracic duct lymph is about 80% that of plasma reflecting the partition of transferrin between plasma and lymph.

Toxins and particles. Foreign substances, both particles and macromolecules, delivered into the interstitial space are transported in lymph. Toxins of large molecular weight such as certain snake venoms are selectively transported by lymph but toxins of lower molecular weight are carried by both blood and lymph (Barnes & Trueta, 1941). Similar considerations apply to toxins absorbed by the intestine (see Chapter 6). Bacteria also are transported by lymphatics (Barnes & Trueta, 1941) and Cole, Petit, Brown & Witte, (1968) have demonstrated the presence of viable bacteria in the thoracic duct of patients during abdominal surgery.

Non-protein constituents
The concentration in lymph of substances which are protein-bound in plasma will be determined by the relative concentrations of the binding protein in blood and lymph. Substances bound to proteins of very large molecular size appear in relatively low concentration in lymph. In human thoracic duct lymph Werner (1966a) found a lymph:plasma bilirubin ratio of 0.83 which roughly reflects the relative concentrations of albumin in plasma and lymph.

Low molecular weight substances of plasma i.e. MW less than 10000 are found in similar concentrations in lymph and plasma. In fasting human subjects, sodium, potassium, calcium and magnesium were found in lower concentration in thoracic duct lymph than in blood while the anions bicarbonate and chloride were in slightly higher concentration. These differences in ion concentration probably reflect differences in the concentration of protein in blood plasma and lymph, the distribution being governed by the Donnan equilibrium. Binding of calcium and magnesium to protein will also affect the distribution. In this same study uric acid and glucose were found to be in slightly higher concentration in lymph than in plasma (ratio 1.11) while creatinine was in lower concentration (ratio 0.94). The explanation of these observations is not clear.

The partial pressure of oxygen in thoracic duct lymph in man and dogs probably mirrors mean splanchnic tissue oxygen tension. It is somewhat higher than that of splanchnic or systemic venous blood and is readily reduced by reduction in splanchnic blood flow and increased by cyanide which inhibits tissue uptake of oxygen (Witte, Cole, Clauss & Dumont, 1968).

Modification of lymph during its transport by lymphatics
The fluid which enters initial lymph vessels is a tissue-modified pericapillary filtrate. The endothelial cells of fine lymphatics probably exert some control over the composition of the fluid in the lymphatic lumen, especially with regard to the amount of particles and solutes of large molecular weight which are taken into these vessels (see above). The question arises of the extent to which this lymph in initial lymphatic vessels is modified during its subsequent transport to the bloodstream.

Possible modifying factors include a filtering action of lymphatic vessel walls, equilibration of lymph with blood plasma in the vasa vasorum of medium and larger lymphatics or when lymph and blood come into close proximity in lymph nodes, and active modification of lymph composition by lymphatic nodes, i.e. by phagocytosis of particles and certain macromolecules and the addition of cells or solutes to lymph. Bulk transfer of lymph to blood through lymphatico-venous anastomoses will not alter lymph composition but will reduce the amounts entering central lymph trunks from peripheral vessels. Much more lymph appears to enter peripheral lymphatics than can be accounted for in thoracic duct flow. For example in rats the lymph from the ileo-colic lymph duct cannulated close to the caecum apparently forms a very large proportion of thoracic duct flow when flow rates in the two vessels are compared, yet occlusion of the ileo-colic duct decreases thoracic duct flow to only a very limited extent (Paldino & Hyman, 1964). A simple explanation for this might involve changes in pressure/flow relationships in the main lymph trunks after occlusion of a major tributary. Alternatively bulk transfer of fluid via lymphatico-venous anastomoses or substantial diffusion of fluid from lymph vessels might play a part. Hargens & Zweifach (1976), on the basis of direct protein analysis of lymph obtained from fine lymphatic vessels of the cat mesentery by micropuncture, have proposed that lymph is concentrated in these vessels by filtration of fluid under hydrostatic pressure while protein is retained within the lymph vessel. In other words, the Starling principle would operate here also. For this concentrating mechanism to work the lymph vessels larger than initial lymphatics must be permeable to fluids, since hydrostatic pressures sufficient to overcome the colloid osmotic pressure of lymph protein are not found in the initial lymphatics.

The mixture of lymph produced as lymphatic trunks from various organs anastomose results in progressive modification of lymph composition. For example, the large hepatic lymph contribution to the thoracic duct will raise the protein concentration of the thoracic duct lymph (see Chapter 8).

As far as large particles are concerned it seems that having once reached the lymph they do not readily leave the vessels by filtration. This conclusion is based on studies where retro-injection of suspensions of particles into lymph vessels fails to cause extravasation of these particles unless lymph vessels are ruptured. Pullinger & Florey (1935) showed that while the fluid component of a suspension of carbon particles injected into a lymph vessel in the mouse ear leaked out of the vessel quite readily the carbon particles themselves were retained within the vessel.

Patterson, Ballard, Wasserman & Mayerson (1958) attempted to assess the extent to which lymph is modified during its transit through large lymph vessels. By infusing radioactively labelled albumin into a leg lymph vessel in the dog and measuring the appearance of the labelled protein in thoracic duct lymph it was found that about ten min after starting the infusion the label appeared in thoracic duct lymph, rapidly reaching a plateau which was maintained throughout the period of infusion and for ten min after the albumin infusion had been stopped and a saline infusion begun (see Fig. 2.4). Dextran of a similar molecular size to albumin behaved in a similar way. Very little of the label in these experiments reached the plasma and the authors concluded that these infused substances of high molecular weight do not leave the lymphatics and are delivered to the circulation primarily by the thoracic duct. They estimated that less than 3% of the infused substances reached the circulation by other routes of which the main one in these experiments was considered to be in the popliteal lymph node. Mayerson (1963) has found that substances of small molecular weight such as sodium, urea and glucose readily leave lymphatic vessels and a dextran of molecular weight 2300 behaves like these smaller molecules. To define the limit of lymphatic permeability with this type of preparation Calnan, Rivero, Fillmore & Mercurius-Taylor (1967) used a range of macromolecules. Lymphatic vessels appeared to be permeable to vitamin B_{12} (molecular weight 1350), cytochrome C (12500) and ribonuclease (14000) but appeared to be impermeable to human growth hormone (21000). Equine myoglobin (17000) appeared to leave the lymphatics to some degree and these workers consider that this is approximately the limit of lymphatic permeability to macromolecules. It is likely that the permeability characteristics of this preparation depend on shape as well as size of macromolecules. These studies give information about large lymph vessels. The morphological appearances of large lymph vessels contrast with those of smaller vessels and this may have some bearing on the fact that large lymph vessels do not allow macromolecules of the size of plasma proteins to filter out. Casley-Smith (1969) has found few or no open junctions and plentiful zonulae adhaerentes and occludentes between the endothelial cells of larger lymph vessels. This contrasts with the appearances of fine lymphatics. Again, the endothelium of large lymph vessels has a well defined basement membrane whereas the basement membrane of fine lymph vessels is frequently deficient.

Despite these morphological considerations, however, it may be that even relatively fine lymph vessels have some ability to retain macromolecules, such as proteins, once these have passed into the lumen of the vessel. Zweifach & Prather (1975) noted that when Evans Blue dye was injected by micropipette

65

into the fine lymphatics of the cat mesentery there was diffuse escape of dye from the vessel into the interstitium but when the dye was complexed with albumin the material was retained within the vessel except for small focal leaks. Theories explaining the uptake of protein and fluid by initial lymph vessels are plentiful but it is difficult to test their validity. These observations by Zweifach & Prather (1975) are of particular interest in that they suggest that once proteins are in the initial lymph vessels they may be trapped and able to exert osmotic pull on tissue fluid, thus establishing flow of lymph in these fine vessels. What mechanism could be responsible for uptake of protein into initial lymphatics, however, is quite unclear but explanations which do not postulate an active transport of this protein are not, in general, satisfactory.

Modification of lymph by lymph nodes

Using the popliteal lymph node preparation (Fig. 3.8), similar to that devised by Drinker, Field & Ward (1934), Mayerson *et al.* (1962) have obtained results that suggest that lymphatico-venous passage of small molecules occurs in lymph nodes. Fig. 3.9 shows the cumulative recovery in nodal lymph and plasma of various substances perfused through the popliteal node of the dog via an afferent lymphatic vessel. It can be seen that virtually all infused albumin is recovered while smaller molecules appear in nodal plasma and are incompletely collected in

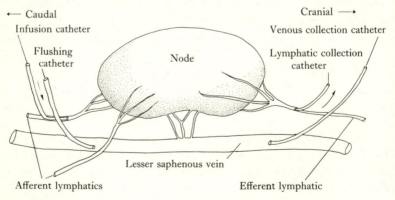

Fig. 3.8. Popliteal node preparation. (From Mayerson *et al.*, 1962.)

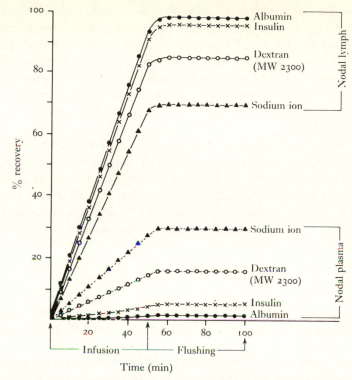

Fig. 3.9. Cumulative recovery of substances of different molecular size perfused through the popliteal node via an afferent lymphatic vessel. (From Mayerson *et al.*, 1962.)

lymph. The transfer of substances of small molecular weight to plasma by this route is presumed to be by a passive process of diffusion. It is, therefore, likely that this equilibration of lymph and blood plasma *in vivo* may be of small importance since the concentrations of the small molecules which are free to diffuse are rather similar in plasma and lymph.

Exchange of fluid and solutes between lymph and plasma in lymph nodes is probably bi-directional. The studies of Quin & Lascelles (1975) indicate that, judged by a correlation between lymphocyte output and plasma protein concentration in nodal efferent lymph after antigenic stimulation, plasma protein is added to lymph during the passage of lymphocytes from blood to lymph through post-capillary venules of lymph

nodes. This may represent bulk flow of plasma in the wake of lymphocytes as they pass through the blood vessels. Peripheral lymph which is derived from tissues containing no lymphoid tissue contains relatively few cells, some in transit from the bloodstream and some derived from the tissues themselves. Passage through a lymph node results in the addition of lymphocytes derived from the node itself or from blood via post-capillary venules.

Taking into account the modifications to lymph brought about by lymph vessels and nodes and the fact that lymph in central channels such as the cisterna chyli and thoracic duct is a composite of various regional lymphs, it is clear that this 'central' lymph, which is the most readily sampled, is unlikely to give much information about the composition of the lymph in initial lymph vessels.

Lymph flow

In general terms, the rate of formation of lymph from an organ or area is determined by the amount of interstitial fluid forming and this in turn is related to blood capillary permeability or perfusion pressure in the capillary bed of the tissue. Increases in any of these will augment lymph flow provided there is no hindrance to the passage of fluid into initial lymph vessels.

Lymph flow rates

As changes in the state of blood microcirculation in tissues alter lymph flow it is not surprising that increased tissue activity of any sort enhances lymph flow. This applies not only to skeletal muscle, where mechanical factors also contribute to the clearance of interstitial fluid, but also to glandular tissue as shown for example by alterations in ovarian lymph flow at different stages in the oestrous cycle (Morris & Sass, 1966). Other factors altering capillary blood flow such as change in local temperature will also affect lymph flow (Ackerman, 1975) and local injury will clearly cause alterations in blood flow and capillary permeability and hence lymph flow (see below). Histamine given intravenously increases thoracic duct lymph flow (Shim, Pollack & Drapnas, 1961) and its local release may be partly responsible for augmented flow from the injured area.

Recently, Deysine, Mader, Rosario & Mandell (1974) have suggested that lymph flow may be influenced by changes in serum calcium levels. In both intact and parathyroidectomised dogs a rise in serum calcium leads to a marked enhancement of thoracic duct lymph flow. The explanation of this finding is not clear. The authors suggest that calcium, by influencing skeletal and smooth muscle activity, might increase lymph propulsion by both extrinsic and intrinsic effects on the lymphatic vessels. Alternatively the rise in lymph flow might be due to an increase in perfusion pressure in blood capillaries since they observed a small rise in aortic blood pressure in response to an increase in serum calcium. Other agents which increase blood pressure, such as vasopressin and angiotensin, increase thoracic duct lymph flow in rats (Fujii & Wernze, 1966).

Thoracic duct lymph flow ranges between 1 and 4 ml per kg per hour for a number of species (Yoffey & Courtice, 1970). For man, the figure is usually estimated as about 1 ml per kg per hour (Crandall *et al.*, 1943; Courtice *et al.*, 1951; Bierman *et al.*, 1953) while for cows the figure is close to 4 ml per kg per hour (Hartmann & Lascelles, 1966). Shannon & Lascelles (1968*a*) give a figure as high as 9.7 ml per kg per hour for calves. For other mammalian species such as dog, cat, rat and rabbit figures of 2–3 ml per kg per hour are found. It should be remembered that flow rate in the thoracic duct cannot be taken as a measure of total lymph formation as the volume of lymph is probably reduced as it passes centrally by discharge of a proportion of the lymph through lymphatico-venous anastomoses to the bloodstream. Yoffey & Courtice (1970) calculate that in non-ruminant animals the volume of lymph passing through the thoracic duct per day is between 50 and 100% of the plasma volume while in ruminants the figures can be as high as twice the plasma volume. Since thoracic duct lymph flow almost certainly underestimates the total rate of formation of lymph in tissues the magnitude of extravascular fluid transport can be appreciated.

Clearly, the flow of intestinal lymph is strongly influenced by digestive and absorptive activity (see Chapter 6). The high figure of 9.7 ml per kg per hour for young milk-fed calves obtained by Shannon & Lascelles (1968a) reflects this influence of digestive activity on lymph flow. All authors agree that intestinal lymph together with hepatic lymph constitutes the major part of the flow.

Propulsion of lymph

Contractility of lymph vessels. Contractility of lymph vessels was described by Hewson (1774) (see Chapter 1) and since then it has been known that fibres of the autonomic nervous system supply contractile elements in the walls of lymph vessels. Medium-sized and larger lymph vessels have considerable amounts of smooth muscle in their walls while the smallest vessels in certain areas, such as the intestinal villus, which consist only of a single endothelial cell layer have a close association with local smooth muscle cells considered by some to confer contractile properties on even these very small vessels (Collan & Kalima, 1970). It is even possible that endothelial cells such as lymphatic endothelium may possess contractile proteins (Pollard & Weihing, 1974) and thus have intrinsic contractility.

Lymph 'hearts' have been long recognised in lower animals. Panizza (1833) described such structures in reptiles and the part they play in returning lymph to the bloodstream. During this century a number of investigators have studied the rhythmic contractility of mammalian lymphatics (Florey, 1927a, b; Carleton & Florey, 1927, and Pullinger & Florey, 1935) and have observed rhythmically propagated contractions of lymph vessels in guinea pigs and rats, while similar activity of lymphatic vessels in the bat's wing was described in 1926 by Carrier and further studied by Webb & Nicol (1944). These same authors (1951) noted that this activity persisted despite denervation of the area. Smith (1949) studying the spontaneous contractility of peripheral lymph vessels in rats, mice and guinea pigs observed that the frequency of contractions increased with the rate of lymph formation and concluded that the lymphatic contractions were stimulated by an increase in intraluminal pressure. In several species lymphatic contractility has been difficult to demonstrate. For example, the canine thoracic duct shows no evidence of spontaneous contractility in either anaesthetised or non-anaesthetised animals (Browse, Lord & Taylor, 1971) though Mawhinney & Roddie (1973) found spontaneous contractions in isolated bovine lymph vessels. The frequency of these contractions was increased by noradrenaline and the effect was abolished by α-receptor blockade with phentolamine. Isoprenaline slowed the spontaneous activity and this effect was counteracted by β-receptor blockade with propranolol. Acetylcholine had little effect on spontaneous contractions and these results were taken to indicate that the autonomic innervation of the thoracic duct in this species is more likely to be adrenergic than cholinergic. Alterations in transmural pressure of isolated bovine lymphatic appear to govern the degree of intrinsic contractility of these vessels (McHale & Roddie, 1976).

The isolated thoracic duct of the dog shows very similar responses and the predominant autonomic receptor appears to be an α-adrenergic receptor. In this preparation acetylcholine did cause smooth muscle contraction but only at relatively high concentration (Tirone, Schiantarelli & Rosati, 1973). These pharmacological studies confirm the early observation by Florey (1927a, b) that lymph vessels of the cat respond to topical adrenaline by contraction. In man intravenous injection of noradrenaline leads to a prompt increase in thoracic duct lymph flow with a subsequent diminution in flow (Leandoer & Lewis, 1970). These authors consider that this response reflects a pharmacological action on the lymph vessels causing constriction and expulsion of pre-formed lymph in the vessels.

There is no doubt that there are wide differences among various species in the degree of spontaneous activity demonstrable in lymph vessels and anaesthetic and surgical procedures are probably able to suppress temporarily a weak contractile mechanism. Hypotension produced by anaesthesia and surgery might be expected to reduce lymph formation to a very low level and further reduce any spontaneous activity of the vessels. There is great variation in the amount of smooth muscle in lymph vessels among different species. Schipp (1967) classified various species into four groups using this criterion as applied to mesenteric lymph vessels. Bats have much smooth

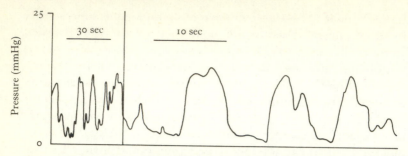

Fig. 3.10. Pressure pattern from an intestinal lymph duct of a sheep with open outflow cannulation of the duct. The rate of lymph flow was 35 ml per hour and the frequency of pulsations was 26 per minute. (From Hall *et al.*, 1965.)

muscle; rats, guinea pigs, monkeys and humans have moderate amounts; cats and squirrels have little, and hamsters virtually none. There is quite good correlation between the amounts of smooth muscle and the ease with which contractility can be demonstrated.

Hall, Morris & Wolley (1965) have made a careful study of rhythmic contractions of several lymphatic vessels of the sheep. These have included the thoracic duct, the intestinal trunk and the hepatic vessel (Fig. 3.10). Regular rhythmic contractions were observed generating pressures of 1–25 mmHg and with frequencies of up to 30 per min. Obstruction produces pressures of up to 60 mmHg and as Smith (1949) had observed, the frequency of contractions rose with the rate of lymph formation. Smeaton, Cole, Simpson-Morgan & Morris (1969) have recorded similar rhythmic contractions in lymph vessels of the foetal lamb *in vivo*. The relationship between lymph flow and contractility of the vessels in non-anaesthetised sheep has been studied by Campbell & Heath (1973) who have compared lymph flow with the pulsatile component of lymphatic pressure and shown that at various flow rates a strong correlation exists. This study used various lymph vessels including the mesenteric lymphatics where augmented lymph flow rates were achieved with fat feeding.

In man, Kinmonth & Taylor (1956) made the first direct observations of spontaneous contractility in lymph vessels during surgical operations. The heat of an operating lamp increased the frequency of contractions which may have resulted from increased lymph formation due to local vaso-dilatation. Kinmonth & Sharpey-Schafer (1959) recorded pressures in normal human thoracic ducts of between 10 and 30 mmHg with increases of 10 mmHg or more during contraction of the vessel, and with obstruction, pressures of up to 50 mmHg, figures similar to those of Hall *et al.* (1965) for the sheep. In human subjects with chronic lymphatic fistulae an intrinsic rhythmicity of approximately 5 contractions per min has been reported by Tilney & Murray (1968).

These several studies indicate that lymphatic vessels of many mammals show a spontaneous rhythmic contractility. The role of the autonomic innervation of small lymph vessels on their contractility is not clear since it is difficult to know whether stimulation or interruption of such nerves produce their effects directly on the lymph vessel or indirectly through alterations in local haemodynamics. Campbell & Heath (1973) consider that, although there are large numbers of autonomic nerves in the walls of lymphatics, they are not necessary for the initiation or maintenance of contractility since contractions will continue for more than 30 min after death. It is particularly interesting that the spontaneous activity of lymph vessels is related to the load, that is, the rate of formation of lymph, and it may be that the stimulus to contraction of the smooth muscle of the vessels is the attainment of a threshold intraluminal pressure. The force and amplitude of a contraction of a lymphangion probably depends on distension of that unit of the lymph vessel just prior to contraction (Mislin, 1976). Thus the mechanism would be analogous to Starling's law of the heart.

There has been little work on the electrical activity of lymphatic smooth muscle. However Azuma, Ohhashi & Sakaguchi (1977), using the sucrose-gap technique, have demonstrated that the resting potential of smooth muscle of bovine mesenteric lymphatics is about 33 mV, a value somewhat less than that generally recorded in vascular smooth muscle. It was also found that an action potential of about 3 sec duration always preceded a contraction wave in the muscle by about 750 millisec.

General physiological considerations

Indirect mechanisms. Lymphatic valves impose a direction on lymph flow and clearly lymph vessel contractions supply a motive force. To what extent do other indirect mechanisms promote lymph flow? The movement of the thoracic cage in respiration undoubtedly plays some part. Hall *et al.* (1965) showed that variations in intrathoracic pressure influence lymph flow in the thoracic duct and considered that the 'thoracic pump' could aspirate lymph towards the thorax from other regions. Transmitted arterial pulsations and movements of skeletal muscles close to lymphatic vessels probably contribute a little towards lymphatic propulsion by direct massaging action. Muscular movements of the leg in adult sheep are minimally transmitted to the lymphatics. The lymph flow from unexercised limbs is normally small and the rise in flow which occurs on exercise is probably due to a large extent to alterations in blood flow to the muscles. Muscular activity, by altering pressures in interstitial spaces by mechanical actions, may propel fluid into lymphatic vessels, a situation which can be simulated by passive movement of the limbs of anaesthetised animals. In the intestine, contraction of the muscle coats would be analogous to skeletal muscle activity and the activity of smooth muscle of the intestinal villi, possibly under humoral control (Kokas & Johnston, 1965), may represent a special mechanism for effective propulsion of the contents of the central lacteal. This 'villus pump' would be an efficient means of promoting rapid flow during absorption especially where large amounts of fluid and lipid are being transported.

Zweifach & Prather (1975) have made a detailed study of the pressures in terminal lymphatics and collecting vessels in the mesentery and omentum of anaesthetised cats, rabbits and rats using micropipettes. In the initial lymph vessels of cats, pressures ranging between zero and 2.6 mmHg were recorded. These authors believe that such pressures probably reflect surrounding tissue pressures though such a contention is difficult to prove; but if this is so the concept of negative interstitial fluid pressure (Guyton, 1963) must be called into question. As terminal lymph vessels converged to form collecting channels, intralymphatic pressures rose progressively in each intervalve

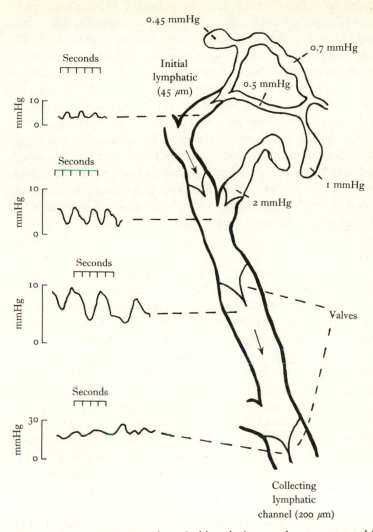

Fig. 3.11. Composite drawing of terminal lymphatic network as reconstructed from several mesentery preparations. The entire network is rarely seen in one preparation. Note increase in intralymphatic pressure in larger collecting channels. (From Zweifach and Prather, 1975.)

75

segment and there was a jump in pressure of roughly 0.75 to 1.5 mmHg across each valve. Using two micropipettes it was possible to show that a pressure increment of 0.75 to 1.2 mmHg would open a valve. The progressive rise in pressure across the valves presumably reflects the competence of the valve, the tone of the smooth muscle in the vessel wall and the relative impermeability of the vessel. Fig. 3.11, composed from results in rat mesentery, illustrates this and further shows pulses of pressure in the larger vessels due to lymphatic contractions. While these data cannot be challenged, it is difficult to understand how lymph flows from a region of low pressure to a high pressure zone, and one must postulate that there is a 'diastolic' interval in the lymphangion during which period pressures fall and flow occurs across the downstream valve.

In a mucosal membrane preparation from the dog small intestine, Lee (1974) has measured the pressure in the central lacteal of the intestinal villus with micropipettes. The values obtained ranged around zero but it was noticed that the pressures fluctuated towards negative values. This could be due to the activity of the contractile elements of the villus, i.e. during villus relaxation the pressure might fall to a negative value thus encouraging fluid to enter the lacteal, and villus contraction would raise the pressure driving lymph forward to draining lymph vessels.

In summary, it seems likely that active propulsion of lymph by intrinsic rhythmic contractions of smooth muscle in the vessel walls is a major element in lymph flow. Skeletal muscle activity possibly adds a supplementary force by its effects on local interstitial fluid pressures, as does the 'thoracic pump'. Lymphatic valves are probably of considerable importance in assisting the process. Lymphatic insufficiency in the form of faults in the intrinsic contractile mechanism or incompetence of the valves may be an important component of some forms of oedema. Cyclical changes in venous pressure in great veins in the neck influence the flow of lymph from the thoracic duct to the venous system (see Chapter 8).

Filling initial lymphatics

The nature of the motive force responsible for filling initial lymphatics is unsettled, though theories abound. In those species which have marked propulsive activity in medium-sized vessels, a suction effect can be envisaged which would be based on a fall in intravascular pressure in 'diastole' and would depend on the competence of the lymphatic valves. It is more difficult to explain the phenomenon in those species where lymphatic contractility is slight or absent. The concept of a negative interstitial fluid pressure (Guyton, 1963) would seem to add to the problem of uptake of fluid into initial lymphatics if the pressures in these vessels are zero or positive as most investigators suggest. The only circumstances, therefore, in which a hydrostatic pressure gradient could drive fluid and solutes into initial lymphatics would be a pathological one where a substantial amount of oedema fluid has collected in the interstitial space. In order to explain the physiological filling of initial lymphatics, Casley-Smith (1972), after discarding various theories involving hydrostatic pressure effects, has proposed that an effective colloid osmotic pressure is responsible. The protein concentration in initial lymphatics is considered to be substantially greater than that in interstitial fluid (Casley-Smith, 1972) and this would create an osmotic pull which exceeds the observed contrary hydrostatic pressure gradient. During tissue compression total tissue pressure rises and is transmitted to the lymph. Ultrafiltration of fluid to the tissues across closed interendothelial junctions concentrates macromolecules by ultrafiltration and dilutes them in the interstitium. Some lymph is driven centrally towards collecting vessels. During a relaxation phase, the concentrated protein in the lymphatics exerts an osmotic pull despite the fact that intercellular junctions have opened. This hypothesis, illustrated in Fig. 3.12, likens initial lymphatics to a host of tiny inefficient pumps which force a small proportion of the fluid which enters them past the first valve into the relatively impermeable collecting lymphatics. This hypothesis is in accord with the views of Guyton, Granger & Taylor (1971) who believe that protein removal from the interstitial space by lymphatics however this is achieved, reduces the colloid os-

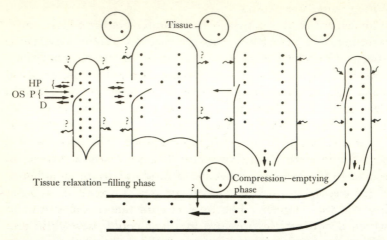

Fig. 3.12. A hypothesis to explain physiological filling of initial lymphatics. The dilation during filling, and compression and telescoping during emptying are shown. The thick arrows indicate protein movement; the thin ones indicate fluid flow. Their lengths show variations in flow rates during the cycles. The wavy arrows indicate the solid-tissue pressure or the pull by the filaments (which may not be important in normal tissue although they are in inflammation). Proteins (dots) are represented in their presumed relative concentrations in the tissues (circles), in the initial lymphatics during the cycle, and in the (thick-walled) collecting lymphatic, before and after their hypothesised dilution by the inflow of fluid through its wall.

During filling, material enters the initial lymphatics via the open junctions. The hydrostatic pressure difference (HP) is shown as either inwards or outwards, according to circumstances, although it is probably only directed inwards in oedema. The hypothesised osmotic pressure difference (OsP) due to the differences in protein concentration causes fluid and proteins to enter the vessels. This pressure difference changes with variations in the concentrations of the intralymphatic proteins. There is also a similarly variable outwards diffusion (D) of proteins. During compression, the junctions become closed; the net hydrostatic-osmotic pressure now forces fluid out of the junctions, leaving the concentrated remaining protein to provide the osmotic pressure needed to cause the cycle to continue. Thus the lymph concentration and volume should be inversely proportional to each other. (From Casley-Smith, 1972.)

motic pressure of the interstitial fluid which is then withdrawn by osmosis into the blood thus creating a negative pressure in the interstitial space.

There is little evidence to substantiate the Casley-Smith hypothesis and recent studies by Rutili & Arfors (1977) provide a strong argument against it. Micropuncture of sub-cutaneous tissue in rabbits has yielded nanolitre samples of interstitial fluid and lymph whose protein concentrations are very similar; this suggests that the endothelial wall of initial lymphatics does not act as an important barrier to the movement of protein either into or out of the lymph vessel. Thus the forces involved in protein and fluid flux from interstitium to lymph vessel are still very far from clear.

Lymph and tissue injury

Lymph flow from inflamed or injured tissue is increased and lymph removes toxic products from these tissues. Since lymph composition in general reflects tissue fluid composition, information can be obtained about the local concentrations of pharmacological mediators of the tissue response to injury. Histamine and components of the kinin system have been found in the lymph draining hind-limbs of dogs after mechanical, ischaemic or thermal injury and prostaglandins of the E series have been found in lymph draining tissue subjected to thermal injury for up to six hours after the injury. Their role in inflammatory responses may involve synergism with the kinins. Thermal injury in the hind-limbs leads to a marked increase in femoral lymph of pre-kallikrein, the zymogenic form of kallikrein which is responsible for the formation of bradykinin. However, 5-hydroxytryptamine, which plays an important role in platelet function in haemostasis, does not appear consistently in lymph draining an injured area, presumably because its action is primarily intravascular. (For a review of this subject, see Lewis, 1975.)

A variety of enzymes, some cytoplasmic, some associated with mitochondria and some derived from lysosomes, appear in lymph draining tissue injured by various means. The pattern of enzyme release varies with the degree of injury and this implies that in mild injury cell permeability may alter allowing escape of intracellular enzymes while extensive cell lysis is required for release of other enzymes (Lewis, 1975). For example, release of lysosomal hydrolases was detected in the lymph only after extensive cell destruction following freezing and thawing of a limb, whereas after mild thermal injury lactic dehydrogenase, glutamic–oxaloacetic transaminase and glutamine pyruvic transaminase escape into lymph.

Although this has not been extensively studied it is likely that these various pharmacologic agents and intracellular enzymes appear in lymph draining areas of the gastro-intestinal tract affected by pathological processes ranging from mild inflammatory lesions such as mild colitis to extensive autolysis of tissue such as acute necrotising pancreatitis or acute ischaemia of a segment of the intestine. It is even possible that the normal gastro-intestinal functions, digestion and absorption of a meal constitute a form of mild physical and chemical injury with release of some products of tissue injury such as histamine or intracellular enzymes into interstitial fluid and lymph.

Summary

The microcirculation allows free exchange of fluid and crystalloids between plasma and tissues. A small proportion of plasma protein escapes into tissue fluid and this requires a means of return to the bloodstream. The extravasated protein and associated fluid returns by the lymphatic system; the importance of the lymphatic system is emphasized by the magnitude of this extravascular circulation of plasma protein (Table 3.1). The forces involved in fluid and solute exchange across the blood capillary wall are well defined and have been measured directly and indirectly. Gastro-intestinal lymph constitutes a major component of thoracic duct flow and is formed from a vascular bed which is relatively permeable to macromolecules. Liver lymph, considered in Chapter 8, is formed from a low pressure vascular bed of even greater porosity. In general the blood flow to any tissue is determined by its physiological activity; while this probably also applies to the intestine there are problems in quantitating the increase in blood flow during digestion and absorption in the gut. The composition of intestinal lymph reflects the permeability characteristics of the intestinal microcirculation and the metabolic and absorptive activity of the gut and is therefore an indication of tissue fluid composition in the gut. However, lymph composition may be altered by the smaller lymph vessels. The forces involved in the entry of fluid, solutes and particles into initial lymph vessels are incompletely understood and the various theories require experimental verification. In the gut, an important contribution comes from the *vis a fronte* generated by propulsion of lymph by smooth muscle activity in the larger lymph trunks which, together with extrinsic forces such as the motor activity of the intestine and the thoracic pump, is responsible for lymph flow to central lymph trunks and the blood. Valves maintain unidirectional flow.

INVESTIGATIVE TECHNIQUES IN
LYMPHATIC PHYSIOLOGY

In describing the general anatomy and physiology of the lymphatic system, reference has already been made to certain techniques used in its study. In Chapter 3, the use of electron microscopy in the investigation of the permeability of lymphatic capillaries to various substances was discussed. An example of such an approach is the ultrastructural study of the penetration of initial vessels in the intestinal mucosa by chylomicra. Mention should also be made of techniques where dyes, and other foreign materials which are selectively taken up by lymph vessels, are injected into the interstitial space permitting visualisation of local lymphatics. This procedure is a necessary preliminary in the technique of lymphangiography in man, prior to direct intralymphatic injection of radio-opaque contrast medium.

Cannulation of lymphatic vessels

The greatest advances in understanding the function of the lymphatics have however come from techniques which involve cannulation of the lymphatic vessels. These permit the establishment of acute or chronic fistulae or, by means of a shunting device, allow intermittent sampling of lymph over a long period. Such methods offer the possibility of studying flows and pressures of lymph and contractility of the vessels, and have been used extensively to investigate capillary permeability to macromolecules. These techniques have also been of particular value in studying lymphatic function in relation to gastro-intestinal physiology in the last twenty-five years.

The availability of fine flexible plastic tubing has made the preparation of chronic lymphatic fistulae, which will function in the conscious animal, a practical possibility. Fine glass

cannulae used before this carried obvious mechanical dis-
advantages and allowed only acute studies under anaesthesia.
Furthermore, lymph is particularly liable to clot on contact
with glass; siliconised glass reduces this problem to some
extent. Many investigators did use glass cannulae to study flow
and composition of lymph from major vessels of animals under
anaesthesia. General anaesthetics, however, appear to alter
flow and composition of lymph, each anaesthetic having its own
particular effect (Polderman, McCarrell & Beecher, 1943;
Hungerford & Reinhardt, 1950).

In 1948, Bollman, Cain & Grindlay described the preparation
in the rat of chronic fistulae of the cisterna chyli, the main
mesenteric lymph vessel, and the hepatic lymph vessel using
fine polyethylene tubing. These preparations have been widely
used in investigating gastro-intestinal physiology. Clear accounts of the technique are to be found in their original paper
and in the appropriate chapters of Lambert's *Surgery of the
Digestive System of the Rat* (1965). Fig. 4.1, taken from this
source, illustrates diagrammatically the anatomical relationships of the intestinal and hepatic lymph vessels of the rat as
they pass towards the cisterna chyli and thoracic duct on the
posterior abdominal wall.

Girardet (1975) has recently described surgical techniques
for long-term studies of thoracic duct circulation in rats. Two
techniques – thoracic duct shunt and thoracic duct side
(T-tube) fistula – allow repetitive sampling while preserving
physiological conditions. In these studies, ten shunts func-
tioned for eight to twenty-six days and ten side fistulae func-
tioned for eight to thirty days. Average lymph flows in the two
preparations were similar, that is, 0.044 ml per min (shunts)
and 0.042 ml per min (side fistulae). The animals, which
weighed between 350 and 570 g, remained in good general
health with minimal loss of lymph and lymphocytes. Thus,
average blood lymphocyte counts, 11700 cells per mm^3 (shunts)
and 12600 cells per mm^3 (side fistulae) remained stable
throughout the study. Thoracic duct lymphocyte transport
calculated from flow rate and lymphocyte count of a sample
also remained stable throughout the study, that is, approximately 1.7 million cells per min (shunts) and approximately 2.3

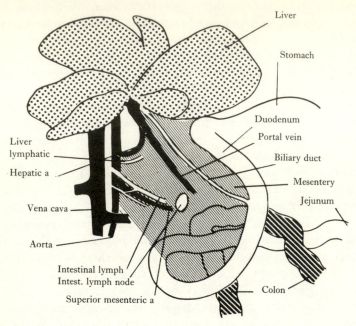

Fig. 4.1. Anatomical relationships of the intestinal and hepatic lymph vessels in the rat (Lambert, 1965.)

million cells per min (side fistulae). The stability of these parameters can be contrasted with the effects of a thoracic duct end fistula in ten rats during seven days of thoracic duct drainage where the blood lymphocyte count and thoracic duct lymphocyte output decreased substantially over a week, as might be expected with the loss of recirculating lymphocytes. The thoracic duct shunt is a difficult technique but the T-tube fistula is no more difficult than the technique described by Bollman *et al.* (1948).

The need for restraint of rats with lymphatic fistulae resulted in the development of a restraining cage (Bollman, 1948) illustrated in Plate 4.1. Most investigators who use chronic lymphatic fistulae in rats drain the lymph to the exterior; Tasker (1951), however, developed an intra-abdominal glass reservoir for collections. Accurately-timed short-term collections are difficult to obtain by this method and it is no longer used.

Technically, the methods described by Bollman *et al.* (1948) require practice and some dexterity. The hepatic lymph vessel presents more formidable difficulties than the others, amongst these being the identification of the vessel, which contains rather clear lymph compared with the turbid contents of in-testinal lymph vessels. Friedman, Byers & Omoto (1956) have given a detailed account of the technique of cannulating this vessel in the rat. Their fistulae were permitted to run for twelve hours. A major problem in preparing lymph fistulae in small animals lies in securing the cannula. The introduction of glues suitable for biological use, which harden in the presence of small amounts of water by a polymerisation reaction, has made these techniques easier (Gallo-Torres & Miller, 1969). Boak & Woodruff (1965) have described the use of such a glue in establishing chronic thoracic duct fistulae in mice.

An important problem associated with lymphatic fistulae is the formation of clot in the cannula. This is seldom encoun-tered later than twenty-four hours after operation in rats. Operative trauma, with the release of small amounts of blood and tissue debris into lymph, can initiate the process. Intragas-tric administration of 0.9% sodium chloride after operation, or the use of a slow intravenous infusion of saline, ensures a good flow of lymph and helps to reduce the risk of clotting (see Chapter 6). Lee & Hashim (1966) have described a technique for preventing coagulation by adding heparin to the lymph by a slow infusion into a mixing chamber built around the draining catheter close to the lymph vessel.

Although the rat has been very widely used in such pre-parations, acute and chronic lymphatic fistulae have been pre-pared in many other species. For example, Shrewsbury (1959), Gesner & Gowans (1962), Boak & Woodruff (1965) and Mandel (1967) describe the technique of thoracic duct cannulation in mice. Thoracic and right lymphatic ducts of rhesus monkeys have been cannulated in the thorax for chronic studies (Hodges & Rhian, 1962). Chronic intestinal lymphatic fistulae have been used in goats for studies of fat absorption (Lascelles, Hardwick, Linzell & Mepham, 1964). The sheep has proved to be a most valuable experimental animal in this connection; the prepar-ation of chronic fistulae of the thoracic duct, hepatic, intestinal and mammary lymph vessels was described by Lascelles & Morris (1961) and by Felinski, Garton, Lough & Phillipson (1964). The former authors describe the return of lymph to the venous circulation in some of their animals. Smeaton, Cole, Simpson-Morgan & Morris (1969) have described a means of preparing chronic lymphatic fistulae in foetal lambs *in utero*. These workers have successfully prepared fistulae of the thoracic, intestinal, lumbar and popliteal lymph ducts of these animals and were able to make satisfactory measurements of flow rates and protein content of lymph in the foetal animal and to demonstrate spontaneous contractions of these vessels. Joel & Sautter (1963) used calves to prepare a thoracic duct–jugular venous shunt which would permit studies of lymph flow and composition over several days.

The lymphatics of the dog have been widely used in experimental studies. Prior to the introduction of plastic tubing for cannulation, several studies were conducted on canine lymph flow under anaesthesia (see Yoffey & Courtice, 1970, chap. 3). Cain *et al.* (1947) using polyvinylchloride tubing reported success in creating thoracic duct fistulae which remained patent for four to eight days. They also cannulated a major liver lymph vessel. Subsequently, Glenn *et al.* (1949) detailed the techniques for creating chronic thoracic duct fistulae in the chest, and fistulae of hepatic, intestinal and iliac lymphatic trunks; they have described the maintenance of these fistulae over several days in the unanaesthetised dog. Many descriptions of modifications of these techniques are available (see, for example, Rampone, 1959, and Viegas de Andrade, Bozymski & Sessions, 1968). Lindsay (1974) has recently suggested that the dorsal saccular portion of the cisterna chyli is a more suitable vessel for collection of lymph in the dog or cat because it is less liable to anatomical variation than the thoracic duct. As with chronic lymph fistulae in all species, metabolic disturbances in the dog are a major consideration (see below). Protein metabolism and fluid balance are notably affected (Glenn *et al.*, 1949). To overcome this problem, Brown & Hardenbergh (1951), Doemling & Steggerda (1960), Nelson & Swan (1969) and Dedo & Ogura (1975) have described the preparation of thoracic duct–venous shunts which allow intermittent sampling of lymph over long periods, and Kolmen (1961) has used a thoracic duct–oesophageal fistula for this purpose. Wang, Caro & Yamazaki (1969) have created an exteriorised lymphatico-venous shunt for hepatic lymph studies in the dog.

Early studies of the human gastro-intestinal lymphatic system made use of rare opportunities such as a spontaneous lymph fistula or chylothorax. Numerous studies have been made on such patients (see, for example, Munk & Rosenstein, 1891; Drummond, Bell & Palmer, 1935; Fernandes, van de Kamer & Weijers, 1955). Surgical approach to studies of human lymph began when Bierman *et al.* (1953) outlined a method for creating a chronic thoracic duct fistula for therapeutic purposes and studied flow, chemical composition and cell count of lymph in patients with haematological disorders. The fistulae were open for periods of two to thirteen days. Intravenous infusions of whole blood or plasma were given to replace the drained lymph; no deleterious effects were observed in the clinical condition of the patients.

Linder & Blomstrand (1958) described a technique of thoracic duct cannulation in man which allowed eventual withdrawal of the cannula and repair of the thoracic duct. They made collections over a few days and stressed the importance of careful measurement of plasma proteins and electrolytes, and the correction of deficits of these substances during the drainage period. Werner (1965) has made an extensive study of the technique of creating chronic human thoracic duct fistulae and the management of such patients. On follow-up, no deleterious effects were found in these patients as a result of the chronic fistulae (Werner, 1966a, b). Fish *et al.* (1969) have kept thoracic duct fistulae open in man for periods of up to 150 days using re-infusion via a tributary of the internal jugular vein. Average daily lymph flows varied between 3.2 and 9.8 l per 1.73 m^2 body surface. Recently the therapeutic possibilities of chronic lymph drainage have been explored in a wide variety of conditions. These will be discussed in Chapters 5 and 8.

Metabolic effects of chronic lymph fistulae

In view of the composition and rate of production of lymph, it would be expected that chronic drainage of a central lymph vessel without replacement might deplete the organism of fluid, the chemical constituents of lymph and lymphocytes. Not only protein would be lost but also a large proportion of absorbed lipids and lipid-soluble substances. In addition to diverting all dietary long-chain triglycerides a fistula of the intestinal lymph vessel or thoracic duct would also drain considerable amounts of lipids of endogenous origin (see Chapter 7). The rapid development of hypoprothrombinaemia in rats with lymph diversion (Mann, Mann & Bollman, 1949) reflects such an effect on vitamin K, which is derived from the diet and from intestinal micro-organisms through coprophagy. Coagulation defects found in animals with chronic lymphatic fistulae are, however, complex. Traber, Haynes, Daily & Kolman (1963) have concluded that deficiencies of factors VIII and XI may also contribute to the effect in the dog.

A chronic thoracic duct fistula constitutes a massive drain on the protein economy of the body. This protein loss would include a significant amount of immunoglobulin normally delivered from the intestine to the blood via lymph (see Chapter 6). In rats, protein concentrations in thoracic duct lymph are approximately 20 to 40% of plasma protein levels. Forker, Chaikoff & Reinhardt (1952) found that in unanaesthetised rats, about two-thirds of the plasma protein leaves the bloodstream per day to be returned by the lymph. For other species figures as high as twice the total plasma protein have been given for lymph passing through the thoracic duct each day (see Chapter 3). Rats drinking 1% sodium chloride solution over a four day period appear to lose more protein through a lymph fistula than animals drinking water (Shrewsbury & Reinhardt, 1952; see Chapter 6). A hepatic lymph fistula in sheep was estimated to result in a loss of 10–15 g of protein per day (Lascelles & Morris, 1961). In dogs with chronic thoracic duct fistulae, protein loss was very marked and after a week serum protein concentrations had fallen by as much as 50% (Glenn *et al.*, 1949). The data for flow rates and protein content of human lymph vary substantially in different reports. On average, a free flowing thoracic duct fistula would be expected to drain about 40–50 g of protein per day. The albumin loss will be more marked than globulin loss in view of the differences in the distribution of protein fractions in plasma and lymph (see Chapter 3). Tilney & Murray (1968) have observed that, despite accurate replacement of lymph protein and fluid, a modest fall in plasma protein levels occurs with chronic lymph drainage in man. A more marked fall in lymph protein concentration accompanied these changes in plasma proteins. No explanation of this difference is obvious.

The increasing use of chronic lymph fistulae in man in the management of patients with various diseases of the liver and pancreas has necessitated careful assessment of the metabolic state of subjects during such drainage. Dumont & Witte (1969) suggest that for short periods of external lymph diversion, patients can make good the metabolic losses by mouth, though intravenous infusions can be used where this fails. Where drainage of large volumes of lymph is undertaken, intravenous administration of albumin, plasma or anticoagulated lymph may be necessary to restore protein balance. It is clear that such a diversion of lymph would also produce a considerable loss of electrolytes. Lymph sodium concentration is rather constant despite fluctuations in lymph flow rate (Barrowman & Roberts, 1967), and a rat with a chronic thoracic duct fistula loses approximately 5 meq of sodium per day. This represents more than twice the total sodium content of the plasma. Most authors who describe the use of chronic lymphatic fistulae without re-infusion comment on progressive weight loss and weakness which is ascribed to protein and electrolyte depletion.

Other solutes, like glucose, will also be lost in considerable amounts, since the concentration of glucose in lymph is comparable to that in plasma (Friedman *et al.*, 1956). Koler & Mann (1951) found that rats with intestinal lymph fistulae lost about 0.5 μg of iron per hour.

Following the establishment of a fistula, the lymphocyte count of the lymph falls progressively (Mann & Higgins, 1950; Gowans, 1957). This fall can be prevented by re-infusion of live lymphocytes intravenously (Gowans, 1957).

Investigative techniques

Interpretation of experiments with chronic lymphatic fistulae

The metabolic deficits described above, which occur when lymph is drained without attempt at replacement, will initiate compensatory changes; these must be considered in experiments which are designed to measure the transport of absorbed nutrients from the gastro-intestinal tract. The principal deficits are clearly of fluid, protein and electrolytes.

Caution must be exercised in interpreting data on lymph flow and composition during the first 24 hours after anaesthesia and surgery. Quin & Shannon (1975) have shown that in sheep lymph flow is depressed and lymph protein concentration elevated during this period (Fig. 4.2). This effect is particularly noticeable in lymph from peripheral tissues such as skin and muscle. Changes in cell count in lymph in the post-operative period are also observed; these are variable and depend on the particular lymph duct cannulated.

Anatomical variations in the lymphatic system among individuals of a species, such as accessory or communicating channels, make it difficult to assess the completeness of collection of lymph from a region. Lymph collected from the thoracic duct, close to the jugulo-subclavian tap, is drained from a large area including the intra-abdominal viscera and the hind-limbs. Gabler & Fosdick (1964) have shown in rats with thoracic duct fistulae in the neck that ligation of the cisterna chyli reduces the flow by approximately 90%, indicating the importance of the contribution from the abdominal organs and hind-limbs. Under resting conditions, the latter contribution is probably small. Contributions also come from viscera like the kidneys which are not connected with the alimentary tract.

O'Morchoe & O'Morchoe (1968) have obtained results in dogs which suggested that as much as 50% of thoracic duct lymph flow may be of renal origin. Since renal lymph is drained by several lymph trunks, it is difficult to make a direct assessment of the renal contribution to the thoracic duct and the technique used in the investigation was the measurement of thoracic duct flow before and after renal artery occlusion. It is, however, difficult to be certain that such a manoeuvre does not produce substantial reflex alterations in haemodynamics of the splanchnic circulation. These authors have re-

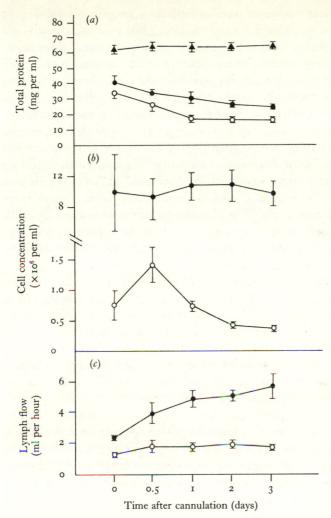

Fig. 4.2. Effect of anaesthesia and surgery on lymph flow and composition in sheep. (*a*) Total protein concentration in afferent (O——O) and efferent (●——●) lymph of the popliteal node, together with the concentration in blood plasma (▲——▲) collected under anaesthesia and at various times following cannulation of the lymphatic ducts. (*b*) Leucocyte concentration in afferent (O——O) and efferent (●——●) lymph of the popliteal node collected under anaesthesia and at various times following cannulation of the lymphatic ducts. (*c*) Flow of afferent (O——O) and efferent (●——●) lymph of the popliteal node collected under anaesthesia and at various times after cannulation of the lymphatic ducts. All values are means of eight sheep±standard errors. (From Quin & Shannon, 1975.)

viewed other studies of renal lymph flow; estimates by different indirect techniques are very variable, but it is clear that renal lymph makes a substantial contribution to thoracic duct flow.

Absorption studies using thoracic duct fistulae in the neck must take into account the possible addition, through hepatic lymph, of metabolic derivatives of the test substance. This same problem arises when the thoracic duct is cannulated below the diaphragm since, in a small animal such as the rat, it may be difficult to be certain whether or not the liver lymph vessel is included in the drainage. Accessory channels, or the opening of collaterals, may result in the inclusion of liver lymph in a situation where this is not desired. Cain *et al.* (1947) have estimated that in the dog hepatic lymph contributed 25–50% of thoracic duct flow and, in view of its high protein content, it must contribute a large proportion of the protein passing through the thoracic duct.

In studies of intestinal absorption it is best to cannulate the main intestinal lymph trunk, tying off the accessory trunk which lies close to it. This eliminates the participation of the liver in the phenomenon which is being examined.

When an attempt is made to define the relative importance of portal venous and lymphatic routes in transport of absorbed substances, the presence of lymphatico-venous anastomoses should be considered. Reference to studies on these anastomoses has been given in Chapter 2. It is probable that communications between intestinal lymph vessels and the portal vein are not normally functioning but exist as potential routes, which open when subjected to increased lymphatic pressure; this may occur when a cannula is partially obstructed or where lymph flows are particularly high.

Other techniques for investigating the function of gastro-intestinal lymphatic vessels

The effect of ligation of lymph vessels on absorptive function has been used in a number of experiments. For example, Annegers (1959) and Planche, Hage, Montet & Sarles (1975) studied lipid absorption, and Murray & Grice (1961) vitamin A absorption in rats following acute lymphatic obstruction.

The ease with which collateral channels open prevents the use of ligation for chronic studies. In acute studies, it is difficult to ensure that such ligation does not cause the opening of lymphatico-venous communications between the absorptive site and the ligature, or even lymphatico-lymphatic channels which bypass the obstruction. Obstruction of the cisterna chyli or thoracic duct may possibly divert hepatic and even intestinal lymph through liver capsular lymphatics, and thence by lymphatic vessels through the diaphragm to alternative intrathoracic lymphatic channels (see Chapter 9).

Lymphangiography is becoming increasingly important as a diagnostic method. Radio-opaque contrast media, injected into subcutaneous lymphatic channels in the feet, are used to visualise major lymphatic trunks and lymph nodes in the abdomen and thorax. At operation local lymphatic vessels may be injected and the drainage channels of a particular area may thus be studied. In dogs Nusbaum, Baum, Rajatapiti & Blakemore (1967) have obtained excellent results in displaying the intestinal lymph vessels at laparotomy by this means; at laparotomy such a procedure in man would be valuable in elucidating the nature of obscure problems which possibly have their basis in lymphatic obstruction. Attempts to visualise intestinal lymph vessels by giving an oral contrast medium have hitherto been unsuccessful.

An interesting approach to the study of intestinal lymph vessels has been made by Lee (1961). In this and subsequent studies he used perfused preparations of small intestine *in vitro* to study fluid and solute transport by cannulated intestinal lymphatics and also measured the pressures within these vessels. These preparations are discussed in detail in chapter 6. More recently, Lee (1969b) has also approached the lymphatic vessels of the intestinal villus directly. A mucosal membrane of dog intestine was mounted on a paraffin block bathed with buffered bicarbonate-Ringer solution and fluid transport was measured in lacteals impaled by glass micropipettes. These various preparations and the experimental results obtained with them are further discussed in Chapter 6. In an organ with no main collecting lymph trunk it is difficult to obtain samples of lymph large enough for analysis. Both the stomach and pancreas, unlike the small intestine or liver, have this disadvantage but microlitre samples of canine gastric lymph have been collected from anaesthetised animals, using glass micropipettes to withdraw fluid from fine lymph vessels on the greater curvature, allowing an analysis of proteins in this lymph and comparison with plasma proteins to be made (Bruggeman, 1975). Papp, Némath & Horváth (1971) have collected pancreatico-duodenal lymph in dogs. Micropipettes have also been used to measure pressures in initial lymph vessels and their collecting vessels (Zweifach & Prather, 1975; see Chapter 3).

5 THE LYMPHATIC SYSTEM AND DIGESTION: GASTRO-INTESTINAL HORMONES AND ENZYMES IN LYMPH

This chapter deals with the part played by the lymphatic system in the digestive functions of the stomach, pancreas and small intestine. The importance of lymph vessels in the subsequent absorption and transport of digested nutrients is discussed in Chapters 6 and 7.

It is possible that one of the roles played by the lymphatics in the process of digestion is the transport of humoral agents which control digestive gland secretion from their cells of origin via lymph to the bloodstream. As far as the digestive secretions themselves are concerned, some observations suggest that certain enzymes derived from the pancreas and small intestinal mucosa appear in lymph in significant amounts and may, by this route, gain access to the blood where they appear in low concentrations under normal circumstances. While the transport of gastro-intestinal hormones by lymph derived from the alimentary tract could be interpreted as a physiological function of these vessels, it is more likely that the presence of digestive enzymes in alimentary tract lymph represents a leakage from their tissues of origin and the lymphatic route for their transport is only of significance in pathological situations. For example, the lymphatics constitute an important pathway for the delivery of large amounts of pancreatic enzymes to the circulation in acute haemorrhagic pancreatitis.

Hormones

Several endocrine secretions are known to be transported in significant amounts by lymph vessels draining the glands where they are synthesised. Examples include thyroid hormones in thyroid lymph (Daniel, Gale & Pratt, 1963) and pressor agents in renal lymph (Lever & Peart, 1962). Humoral agents in lymph draining from the gastro-intestinal tract have been sought for many years. Edkins (1906) suggested the possibility that, after a meal, gastrin might be found in gastric lymph. In 1929, Feng, Hou & Lim, in a study of the inhibitory effect of fat on gastric secretion, were unable to demonstrate a humoral inhibitor of gastric secretion in thoracic duct lymph of dogs given a fat meal, nor could Kim & Ivy (1933) demonstrate any gastric secretagogues in thoracic duct lymph collected from anaesthetised or unanaesthetised dogs over a period of three to four hours after feeding a meat meal. In recent years, however, it has become apparent that lymph derived from the abdominal region does contain substances both stimulatory and inhibitory to gastric secretion. For example, Johnston & Code (1960) found that, during various phases of gastric secretion, the thoracic duct lymph of dogs sometimes contains such inhibitory agents. In a smaller number of cases stimulants of gastric secretion were found in the lymph. These effects were observed when the lymph samples were given by intravenous injection to dogs with vagally innervated or denervated gastric pouches. The greatest degree of inhibition was found with lymph collected immediately after a meal, and the inhibitory agent was found to be present in the non-lipid fraction of lymph. In studies of humoral stimulants and inhibitors of gastric secretion in lymph, identification of specific hormones is essential as certain non-hormone substances present in lymph, such as lecithin, have been proposed as gastric stimulants (MacIntosh & Krueger, 1938).

Kelly, Ikard, Nyhus & Harkins (1963) collected fasting and post-prandial thoracic duct lymph from dogs with Heidenhain pouches and found that the post-prandial lymph injected intravenously into the same dogs stimulated pouch secretion while the fasting lymph did not (Fig. 5.1). It can be seen that considerably more secretagogue was present in the two to four hour post-prandial lymph collection than in the lymph collected in the first two hours. The reason for this is unclear. Ultracentrifugation of the lymph produced a clear subnatant and fatty supernatant, both of which stimulated secretion. Antrectomy resulted in the disappearance of the secretagogue from the lymph. The histamine content of the post-prandial

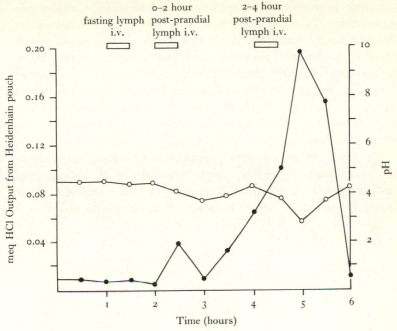

Fig. 5.1. Effect of fasting and post-prandial thoracic duct lymph collected following a liver meal on Heidenhain pouch acid output, ●———●; and pH, ○———○; in a dog. (Kelly *et al.*, 1963.)

lymph was low; clearly it was not the secretagogue involved.

Although it can be seen in Fig. 5.1 that in the particular animal studied some hydrochloric acid secretion from the Heidenhain pouch was initiated before the 2–4 hour sample of lymph was infused, the collected data in this study showed that in six animals basal acid output from the pouch was 0.03 mequiv. hydrochloric acid per 30 minutes. Intravenous infusion of fasting lymph produced no statistically significant increase in hydrochloric acid output but infusion of lymph collected from 0–2 hours and 2–4 hours post-prandially resulted in statistically significant rises to 0.06 mequiv. and 0.25 mequiv. of hydrochloric acid per 30 minutes respectively. These experiments suggest that significant amounts of a humoral agent are present in post-prandial lymph and that this agent is derived from the pyloric antrum. They do not establish whether the hormone is primarily released into the lymph or whether it is secondarily derived from plasma by filtration. Using anaesthetised dogs, Yakimets & Bondar (1966, 1967) studied the effect of diversion of thoracic duct lymph on acid output from Heidenhain pouches, showing that the acid output in response to feeding was greatly reduced by the diversion of thoracic duct lymph. As had been found by Kelly *et al.* (1963), the histamine content of the lymph was not sufficient to account for the observed effect.

The importance of lymph diversion in altering gastric secretory responses to feeding or to the release of endogenous gastrin has been assessed by Rudick, Fletcher, Dreiling & Kark (1968) in conscious dogs. Thoracic duct lymph diversion was

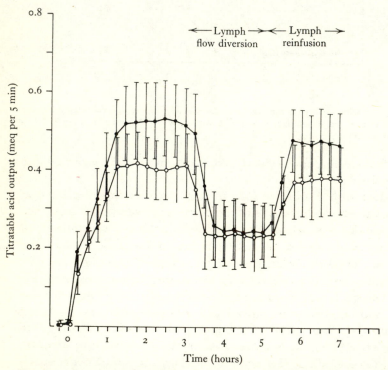

Fig. 5.2. Acid output from Heidenhain pouches of two dogs in response to antral stimulation by 0.5% acetylcholine perfusion. Each point shows the mean±standard error of the mean for three experiments. Once a plateau of secretion was reached lymph flow was diverted for two hours and then restored. (Rudick *et al.*, 1968.)

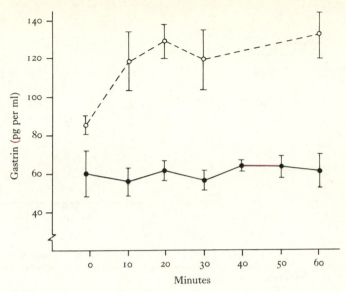

Fig. 5.3. Mean concentration of gastrin (±standard error of the mean) in the serum and thoracic duct lymph of four dogs after feeding. (Clendinnen *et al.*, 1973.)

found to reduce the gastric secretory response to feeding by about 60% and restoration of lymph flow resulted in a return to normal secretion rates. Rudick, Fletcher & Dreiling (1968), followed the secretion of a fundic Heidenhain pouch when gastrin release was induced by irrigation of the antrum with 0.5% acetylcholine. Thoracic duct lymph diversion in such a preparation reduced the secretory response by about a half and restoration of lymph flow resulted in a return to the secretory levels obtained before lymph diversion (Fig. 5.2). Such studies point to an important degree of participation of the lymphatic system in the humoral mechanism of gastric secretion. The work of Clendinnen, Reeder & Thompson (1973), however, has produced results which conflict with those of the experiments quoted above. In their studies of dogs with antral pouches, gastric fistulae and thoracic duct cannulae, gastric acid output was not significantly reduced by lymph diversion and only small amounts of gastrin, as measured by radio-immunoassay, were found to be transported by thoracic duct lymph (Fig. 5.3). Their experiments also included studies of two human subjects

99

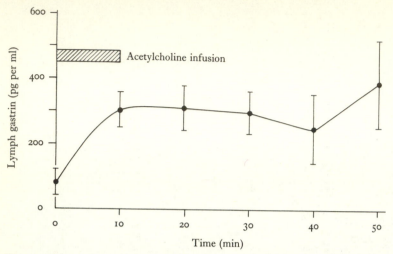

Fig. 5.4. Mean peripheral venous serum gastrin concentrations during and following 10 min infusion of acetylcholine in four dogs. Vertical bars represent ±standard error of the mean. (McGuigan *et al.*, 1970.)

with thoracic duct fistulae, and it was found that the concentration of gastrin in thoracic duct lymph after feeding did not increase, although serum gastrin levels did.

McGuigan, Jaffe & Newton (1970) have followed the release of endogenous gastrin, as measured by radio-immunoassay, from the canine stomach after irrigation of the antrum with acetylcholine. The hormone was assayed in portal, hepatic and peripheral venous blood and in thoracic duct lymph. Gastrin levels in serum were observed to rise in the portal, hepatic and femoral veins in response to antral stimulation. Fig. 5.4 shows the rise in peripheral venous serum gastrin following acetyl-choline irrigation of the antrum. Fig. 5.5 shows that a rise in thoracic duct lymph gastrin also occurred in response to acetylcholine infusion. The levels, however, were lower than those in peripheral venous serum and rose somewhat more slowly. When ^{125}I-labelled human gastrin is given by intra-venous injection to dogs, it appears to distribute throughout the extracellular space (Jaffe & Newton, 1969). In view of these facts and with regard to the relatively slow flow of lymph in the thoracic duct compared with portal blood flow, it would appear that only a very minor amount of endogenous gastrin,

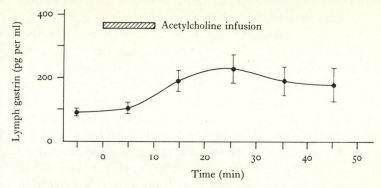

Fig. 5.5. Mean thoracic duct lymph concentrations before and following 10 min infusion of acetylcholine in the same four dogs as in Fig. 5.4. Vertical bars represent ± standard error of the mean). (McGuigan *et al.*, 1970.)

as measured by radio-immunoassay, is delivered to the circulation by the lymphatic route.

It can be seen that there is conflict in the evidence for lymphatic participation in the mediation of the humoral phase of gastric secretion. Certain studies appear to point not only to the existence of gastrin or a gastrin-like secretagogue in post-prandial lymph but also to their presence in amounts which are of quantitative importance in the normal gastric secretory response to feeding. From the data available, it seems likely that gastrin leaves the antrum in gastric venous blood and possibly lymph. The distribution of the hormone between the two routes in the gastric antral mucosa is unknown. The gastrin molecule, 'little' gastrin, a heptadecapeptide isolated by Gregory *et al.* (1964), has a molecular weight of 2114. Larger molecular forms, such as 'big' gastrin, consisting of thirty-four amino acids (see Yalow, 1975) are found in the circulation. The large form, or smaller forms complexed with proteins, might be preferentially taken up by lymphatics. The observations of Kelly *et al.* (1963), Yakimets & Bondar (1966, 1967) and Rudick, Fletcher, Dreiling & Kark (1968) suggest an important role for lymph in gastrin transport, certain experiments indicating that half of the gastric secretory response depends on the lymphatic transport of the hormone. More recent data from McGuigan *et al.* (1970) and Clendinnen *et al.* (1973), however, suggest that negligible contribution is made to the humoral phase of gastric acid secretion by lymphatic transport of secretagogues.

It should be emphasised that most of these studies employ diversion of thoracic duct lymph; this lymph draws from a wide area and includes a hepatic contribution. An assessment of the amounts of secretagogue initially liberated into lymph must await refined techniques such as those described by Aune (1968) and Bruggeman (1975) for collection and analysis of *gastric* lymph together with identification and quantitation of all humoral secretagogues present in this lymph and gastric venous blood. In this connection, it is interesting that pepsinogen-like activity can be found in gastric lymph but is not detectable in thoracic duct lymph (see below).

Reynolds (1967) has proposed that submucosal lymphatic channels connecting the antrum and body of the stomach might convey gastrin to its target, the parietal cells. Injections of dye and India ink were used to demonstrate this anatomical communication in the dog stomach. After alkaline irrigation of the antrum, a submucosal fluid was collected which gave a brisk acid output from a Heidenhain pouch of another animal when given by injection directly into the bloodstream. Few conclusions can be drawn from this data since the secretagogue was not identified, but the intriguing possibility of such a local control of gastric secretion merits further investigation.

Stassoff (1914) found that extensive resection of the small intestine in dogs resulted in gastric hypersecretion. The mechanism of this effect has subsequently been regarded as being humoral in nature since Landor & Baker (1964) demonstrated the hypersecretion in dogs with Heidenhain pouches. In man, a hypergastrinaemia is found in the fasting and post-prandial state after small intestinal resection (Straus, Gerson & Yalow, 1974). The cause of the hypergastrinaemia is debated. Augmented release, possibly through the removal of an inhibitor of gastrin release derived from the small intestine (Straus *et al.*, 1974), or diminished catabolism of gastrin (McGuigan, Landor & Wickbom, 1974) have been proposed. Yakimets & Bondar (1969) have studied the secretion by Heidenhain pouches, in anaesthetised dogs before and after an 80% small intestinal resection in which 60 cm of proximal jejunum were anastomosed to 30 cm of distal ileum. The hypersecretion following small intestinal resection was reduced by more than half by thoracic duct lymph diversion and restored by the re-infusion of the diverted lymph. These data throw no light on the mechanism of gastric hypersecretion after small intestinal resection and can only be interpreted as indicating that a secretagogue, possibly gastrin, is circulating in significant amounts in lymph in this situation.

In addition to gastric secretagogues, it has been proposed that abdominal lymph contains agents which inhibit gastric secretion (Johnston & Code, 1960). These inhibitory agents in lymph have been studied in detail by Rudick *et al.* (1965, 1966). An inhibitor of gastric acid secretion, 'gastrone', has been demonstrated in the gastric juice of man and dogs by several workers (see Rosenthal & Rudick, 1971). Semb *et al.* (1964) showed the presence of this agent in canine antral secretion and found that it has an inhibitory action on histamine- or gastrin-induced gastric secretion. In the experiments of Rudick *et al.* (1965, 1966), dogs with Heidenhain pouches were given infusions of exogenous stimulants, such as histamine. Thoracic duct lymph collected from fasting dogs was found to inhibit the established secretion in the test animals. The inhibitory agent appeared to be neither secretin nor cholecystokinin. In samples of lymph taken after feeding, secretagogues predominated in lymph samples and overcame the action of any inhibitory agents present.

More recently, Rosenthal & Rudick (1971) have partially purified a powerful inhibitor of gastric acid secretion, *chylogastrone*, from canine and human thoracic duct lymph. The lymph was collected from fasting subjects and animals and subjected to chloroform/methanol extraction and gel filtration. The *chylogastrone* preparation, which could only be obtained in the fasting state, appeared to inhibit primarily the secretion of the hydrochloric acid component of gastric juice, and a reduction of pepsin output appeared to be due only to the reduced volumes of juice secreted. A physiological role for such an inhibitor has still to be found, and the chemical nature of this material in fasting lymph is not yet understood.

The demonstration in lymph of humoral agents which

regulate gastric secretion raises the question of whether other gastro-intestinal hormones are carried to an important extent in lymph. Svatŏs, Bartŏs & Brzek (1964) studied the appearance of cholecystokinin, as measured by a bioassay using guinea pig gall bladder, in human thoracic duct lymph and serum following intraduodenal administration of sorbitol. Fasting levels of the hormone in lymph and serum are very low; following sorbitol administration, a sudden rise in the hormone (by about 0.25 Ivy units per ml over basal values) was observed after 10 min in both serum and lymph. Thereafter, the concentrations fell in parallel to basal values in two hours. The results of this study do not indicate that the lymph is of primary importance in the transport of this hormone. However, further assessment of the importance of lymph in transport of this and other digestive tract hormones must rest on the development of sensitive and specific assays of the hormones. Such techniques should allow identification of specific secretagogues in lymph derived from the gastro-intestinal tract and permit quantitative assessment of the relative importance of the vascular and lymphatic routes in their transport.

Insulin in lymph

In recent years it has become apparent that insulin influences secretory and digestive mechanisms in the gastro-intestinal tract. For example, Palla, Ben Abdeljlil & Desnuelle (1968) have obtained evidence which suggests that insulin may control the biosynthesis of pancreatic amylase and chymotrypsinogen. Some comment, therefore, should be made about its presence in lymph and possible lymphatic transport. Since the liver destroys 40–50% of insulin passing in the portal vein (Mortimore & Tietze, 1959), primary lymphatic transport of a proportion of insulin secreted by the pancreas would be of considerable significance as the lymphatics bypass the liver.

Insulin has been measured in thoracic duct lymph of rats (Rasio, Soeldner & Cahill, 1965), rabbits (Daniel & Henderson, 1966), man (Rasio, Hampers, Soeldner & Cahill, 1967) and dogs (Pepin *et al.*, 1970). The presence of insulin in lymph might be due to a direct lymphatic transport of the hormone from the gland or to secondary filtration from the blood, either in the systemic circulation or in the liver. This problem recurs repeatedly when the transport of substances by lymph is considered. Where such substances, either hormones or enzymes, are derived from the pancreas, the problem is not easy because of the difficulties of sampling pancreatic venous blood and lymph.

Histological studies indicate that lymphatic vessels ramify throughout the pancreas in close relationship with acinar tissue but do not penetrate the Islets of Langerhans (Bartels, 1909; see Chapter 2). In rats given intravenous glucose, no rapid rise was observed in the concentration of insulin in lymph in the cisterna chyli, which should drain pancreatic lymph directly (Rasio *et al.*, 1965). Such an observation suggests that no great amounts of insulin are delivered to the circulation by the lymphatic route. Pepin *et al.* (1970) reached a similar conclusion on the basis of comparison of thoracic duct lymph and serum levels of insulin secreted in response to intravenous infusion of glucose or glucagon. As with other materials transported partly in lymph and partly in portal venous blood, the difference between portal venous blood flow rate and thoracic duct lymph flow rate is so great that for quantitatively important lymphatic transport, the concentration of the substance in lymph would have to be much greater than in blood.

The balance of evidence suggests that insulin is principally secreted directly into the bloodstream rather than into pancreatic lymphatic vessels. Insulin in body fluids, however, does not exist entirely as the monomolecular form of molecular weight 5800, but a proportion is present in a much larger molecular form (see the review by Berson & Yalow, 1966). In the macromolecular state, insulin leaves the vascular compartment with difficulty (Rasio *et al.*, 1967). It is possible that if macromolecular types of insulin form in the interstitial fluid, these would be preferentially absorbed by local lymph vessels.

Enzymes

In view of the existence of an extravascular circulation of proteins, it is to be expected that the enzymes of blood plasma should be found in lymph. In human thoracic duct lymph,

Linder & Blomstrand (1958) measured the content of aldolase, alkaline phosphatase, acid phosphatase, transaminase and lactic dehydrogenase, and compared their concentrations as a proportion of total protein with those in serum. Aldolase and acid phosphatase levels, by this criterion, were found to be significantly higher in lymph than in serum, and these authors suggested that the lymphatic route may be a means of delivery of these enzymes to the circulation.

Werner (1966b) has confirmed the finding that in human thoracic duct lymph, aldolase levels are approximately twice those of blood. Most other enzymes are present in lymph in concentrations comparable with or lower than, those of blood. In both man (Dahlbäck, Hansson, Tibbling & Tryding, 1968) and cats, dogs and rabbits (Carlsten, 1950), diamine oxidase (histaminase) is present in lymph in concentrations greatly exceeding those of blood plasma. This enzyme, which originates largely from the kidneys and gastro-intestinal tract, appears to enter the blood by the lymphatic route. In anaphylactic shock in many species plasma diamine oxidase rises; for example, Logan (1961) has demonstrated this phenomenon in guinea pigs. Heparin is also released during anaphylaxis (Jaques & Waters, 1941; Monkhouse, Fidlar & Barlow, 1952). In rats, diamine oxidase is released into the circulation by heparin injection. The major source of the enzyme in this case is the small intestine (Kobayashi, Kupelian & Maudsley, 1969). Dahlbäck *et al.* (1968) have shown in human subjects that injection of heparin results in a brisk rise in diamine oxidase in blood and, from measurements of the enzyme in thoracic duct lymph, concluded that the effect was partially mediated by the lymphatics. It is interesting that fat feeding in rats releases diamine oxidase into intestinal lymph (see below).

In this section, it is proposed to discuss the transport of digestive enzymes in lymph and the significance of their presence therein. In addition to gastric and pancreatic enzymes, alkaline phosphatase of intestinal mucosa is considered. The function of this enzyme is unclear; it does not appear to be primarily concerned with digestion in the intestinal lumen, but rather with absorption or transcellular transport of fat.

Gastric enzymes

Micropuncture of canine gastric lymphatics has shown pepsin-like activity in this lymph after treatment with 100 mM hydrochloric acid. No such activity could be detected on treating plasma, thoracic duct lymph or intestinal lymph with acid (Bruggeman, 1975). This suggests that lymphatics close to secretory epithelium may carry some enzymes from their cells of origin to blood, but the rapid mixture of lymph from other sources reduces their concentration in major lymph trunks.

Pancreatic enzymes

Under normal circumstances, the concentrations of pancreatic enzymes in the blood are low. The rate of addition of these enzymes to the blood is probably slow, though in acute necrosis of the pancreas they are liberated into the circulation at a very much higher rate.

In 1937, Blalock, Robinson, Cunningham & Gray showed that obstruction of the cisterna chyli of dogs led to a massive lymphoedema of the pancreas, thus demonstrating the importance of the lymphatic route in the drainage of the interstitial space of the pancreas. Papp, Németh & Horváth (1971) have shown that the pH of pancreatico-duodenal lymph of dogs is higher than arterial blood and since the pancreatic tissue is relatively alkaline, it is possible that bicarbonate from the duct system of the gland escapes through the interstitium of the gland to the lymph. The question arises as to whether or not a direct functional pathway exists in the normal pancreas by which pancreatic enzymes gain access to the interstitium of the gland and are subsequently drained by lymphatic vessels.

Since the time of Claude Bernard and Magendie, amylase has been identified in blood and lymph, and Carlson & Luckhardt (1908) made a systematic study of amylase concentrations in blood, lymph and other body fluids in various species. In general, they found that serum amylase levels exceeded those of thoracic duct lymph. In their experiments, pancreatectomy in cats did not reduce the concentration of blood and lymph amylase, which indicated that non-pancreatic sources of amylase make important contributions to serum concentrations of this enzyme in this species; though a more recent study by

Singh *et al.* (1969) has suggested that the liver and small intestine in the dog do not contribute to serum amylase or lipase via the lymph. In 1950, Flock & Bollman, using chronic intestinal lymph fistulae in rats, studied the output of amylase and tributyrinase in lymph in fasting and fed states; the lymph in their preparation included contributions from the pancreas. In fasting rats the levels of tributyrinase and amylase in lymph were found to be lower than those in serum. However, feeding a fat-rich meal caused an increase in the twenty-four hour output of both enzymes, especially tributyrinase, from the fistula. While the rise in amylase output was considered to be mainly due to increased total lymph flow, the large tributyrinase output was thought to represent a mobilisation of the enzyme in response to a fat-rich diet. It is difficult to identify the tributyrinase, though it may have been pancreatic lipase.

The direct transport of pancreatic enzymes from the interstitial space of the gland by the lymphatics has been proposed by several groups of workers (Osato, 1921; Dumont, Doubilet & Mulholland, 1960; Duprez *et al.*, 1963; Bartos, Brzek, Groh & Keller, 1966). Early important observations on this subject were made by Osato (1921) who demonstrated in the dog that pilocarpine administration resulted in a large rise in amylase and to a lesser extent in lipase and proteases in thoracic duct lymph, and that this phenomenon was abolished by pancreatectomy. Dumont *et al.* (1960) describe the accumulation of a peritoneal fluid rich in enzymes which appears at splenectomy and is considered to be related to the sectioning of pancreatic lymph vessels at the splenic hilus. In human subjects, these workers have studied the levels of amylase and lipase in thoracic duct lymph, both with the pancreas 'resting' and after stimulation. The administration of secretin alone or in combination with morphine, increased the enzyme concentrations in thoracic duct lymph (Fig. 5.6). Similar studies have been carried out in man by Bartos *et al.* (1966). In their study, stimulation of the pancreas by combinations of secretin and pancreozymin or secretin and morphine caused a rise in levels of amylase and lipase in lymph above normal resting values. Under these circumstances, the levels of enzyme in the lymph exceeded

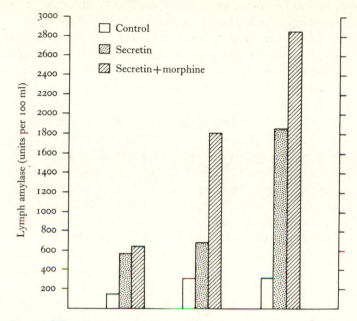

Fig. 5.6. Effect of secretin and secretin plus morphine on thoracic duct lymph amylase concentration in three subjects. (Dumont *et al.*, 1960.)

those of serum and the diversion of thoracic duct lymph markedly reduced the elevation of enzymes in serum which occurred in response to pancreatic stimulation. In anaesthetised dogs with thoracic duct fistulae, the administration of intravenous secretin produced a rise in the concentration of amylase and lipase of lymph (Vega, Appert & Howard, 1967). Before the administration of the hormone, the concentrations of amylase and lipase were in some cases greater and in others less than those of plasma, but after secretin administration, lymph levels were generally observed to be higher. This rise in pancreatic enzyme levels in serum and lymph, which occurs in response to intravenous administration of secretin and pancreozymin, is abolished by pancreatectomy (Singh *et al.*, 1969). Papp, Ormai, Horváth & Fodor (1971) have shown a marked rise in the concentration of lipase in pancreatico-duodenal lymph in dogs in response to infusion of secretin and pancreozymin into the superior pancreatico-duodenal artery.

These various experiments point to the lymphatics as a drainage channel which gains importance during high rates of pancreatic secretion, and when intraductal pressure in the gland rises as happens with, for example, contraction of the sphincter of Oddi brought about by morphine administration. Raised pressure within the pancreatic duct system is known to enhance the transport of pancreatic enzymes to the blood. Gibbs & Ivy (1951) found that ligation of the pancreatic ducts of the dog caused a gradual rise in serum amylase. The rate of rise was greatly enhanced when a pressure of 23 mmHg was applied to the duct system. These observations suggest that the route by which pancreatic enzymes may gain access to the circulation, through either venous or lymphatic drainage of the pancreas, may involve initial passage of enzymes between duct or acinar cells into the interstitial space of the gland. There have been several studies of the pressures required to permit passage of pancreatic duct contents into the interstitial space of the pancreas (see, for example, Herring & Simpson, 1909; Rich & Duff, 1936; Egdahl, 1958). Pirola & Davis (1970) have demonstrated that India ink introduced into the pancreatic ducts of the cat can escape into the interstitium at pressures of 30 and 40 mmHg, pressures considered to be less than the maximum secretory pressure of the pancreas. Edlund, Ekholm & Zelander (1963) have reported interesting electron microscopic evidence for channels between acinar cells which would allow the passage of pancreatic juice into the interstitial space. The evocative test of pancreatic function described by Burton *et al.* (1960) is based on the observation that an adequate but sub-maximal stimulus to the pancreas with secretin and pancreozymin will not lead to a rise in pancreatic enzymes in the blood when the duct system of the pancreas is unobstructed. However in the presence of duct obstruction a rise occurs. Controlled intraductal infusion in the canine pancreas leads to the passage of large quantities of fluid and solute into the interstitium of the gland when the pressure exceeds 30 mmHg (Schiller *et al.*, 1972). Using solutions of fluorescein, it was shown that the bulk of this fluid is cleared by the bloodstream and that the lymphatics of the gland play a secondary role in this clearance. However, concentrations of fluorescein in the thoracic duct were greater than those in peripheral blood and when the pancreatic veins were ligated the lymphatics assumed the major part in clearing the pancreatic oedema, as judged by an increased flow of lymph containing increased concentrations of fluorescein.

It is difficult to reach a conclusion about the importance of lymphatic transport of digestive enzymes from the interstitial space of the pancreas to the circulation under normal conditions. Small amounts of pancreatic enzymes do reach the systemic circulation under normal circumstances, and the lymphatic route accounts for an unknown proportion of the transport. When pancreatic intraductal pressure rises, enzymes gain access to the interstitial space of the gland and are drained, at least in part, by lymph channels. Any situation in which pancreatic venous drainage is impaired, for example by intravascular coagulation in pancreatic veins or by portal hypertension, will enhance the relative importance of the lymphatic route in draining the pancreatic interstitium. One could view blood and lymph drainage of pancreatic juice as a 'safety valve' protecting the acinar cells from deleterious effects of raised intraductal pressure, and therefore a mechanism by which the pancreas is protected from autodigestion by its own secretory products. Most of the studies on this subject involve the use of thoracic duct fistulae, which carries the disadvantage that the lymph is derived from the entire splanchnic area and has contributions from all abdominal viscera. Elaborate indirect experiments are required to establish that changes in the concentration of enzymes, such as amylase, in the thoracic duct lymph are not due to alterations in contributions from sources such as the liver and small intestine. To elucidate the role of lymph drainage in this problem direct access to pancreatic lymphatic vessels is desirable but this presents formidable technical problems.

Acute pancreatitis
In this condition the pancreas becomes the seat of necrosis, haemorrhage, inflammation and oedema. Thrombosis is common in small blood vessels and a secondary bacterial infection of the tissue may supervene. An acute autolytic

process occurs in the gland due to the action of digestive enzymes liberated from the exocrine tissue of the pancreas. The liberation of these enzymes is reflected in a dramatic rise in their concentration in the blood. Hydrolysis of fat due to liberated pancreatic lipase occurs in sites both close to the pancreas and in other areas in the peritoneal cavity such as the omentum. In some cases, disseminated areas of fat necrosis are observed in sites distant from the pancreas, such as bone marrow and subcutaneous tissue. In addition to enzymes, other substances which are produced by enzymic action are released from the necrotic gland. Many of these have potent actions on the circulatory system akin to those of bradykinin, and these toxic substances are considered to be responsible for the severe hypotension which accompanies the condition.

The blood vascular and lymphatic routes are both involved in the drainage of fluid and enzymes, and their digestion products from the pancreas and abdominal cavity. Cells, such as erythrocytes, pass into lymphatic vessels and heavy blood staining of thoracic duct lymph is frequent. In 1947, Perry studied the anatomical pattern of distribution of areas of fat necrosis in rats which were given intraperitoneal injections of a crude pancreatic extract together with finely divided graphite. The lipid necroses were observed to be closely associated with lymphatic channels outlined by graphite, and he concluded that the lymphatic route is very important in the transport of lipase in disseminated pancreatic fat necrosis. This type of experiment does not justify the conclusion that this is the only route for lipase transport after its liberation from the cells or duct system of the pancreas. There is liberal lymphatic drainage of the peritoneal cavity and uptake of foreign materials by these vessels, particularly diaphragmatic lymphatics, is well recognised. It is very probable that the lipase of such a pancreatic extract would be present as a macromolecular complex with a molecular weight of several hundred thousand. In extracts of pancreatic tissue, lipase is generally present in such a state (see, for example, Sarda, Maylié, Roger & Desnuelle, 1964). In macromolecular form, the enzyme might be expected to pass preferentially into lymphatics rather than blood vessels.

In the naturally-occurring pathological process, acute pancreatitis, the areas of disseminated fat necrosis are anatomically related to lymphatic channels. As in the experimental situation, pancreatic lipase probably forms macromolecular complexes with tissue lipids and would thus be directed initially into lymphatic vessels. Erlanson & Borgström (1970) have shown that such large aggregates can form between lipase and lipids of biliary and dietary origin when concentrations of bile acids in samples of intestinal content are artificially reduced by passing these samples through gel filtration columns.

It is conceivable that some other pancreatic enzymes may also form macromolecular complexes or aggregates with tissue components during autolysis of the gland. For example, Ohlsson (1971) has shown that in experimental acute pancreatitis in dogs, trypsin rapidly forms complexes with protease inhibitors of the α_1-antitrypsin type. The complexes appeared in ascitic fluid and, later in the course of the disease, in the thoracic duct lymph. Such complexes are probably taken up by reticuloendothelial cells of liver and lymph nodes, and the delay in appearance of trypsin–macroglobulin complexes in lymph might indicate a saturation of lymph node uptake of these complexes which had efficiently cleared the lymph in the early stages of the disease.

For a quantitative assessment of the importance of blood and lymphatic routes in the dissemination of enzymes and their products, it is necessary to consider transport per unit time rather than the relationship between concentrations in portal venous plasma and abdominal lymph, since even the relatively high flow rates of lymph, which are found particularly in the early phases of the disease, are not to be compared with the portal blood flow. A quantitatively significant transport by the lymphatic route clearly requires relatively high concentration of the substance in question in the lymph. Popper & Necheles (1940), studying anaesthetised dogs with experimental acute pancreatitis, concluded from measurements of amylase and lipase in peripheral venous blood, portal venous blood and thoracic duct lymph that, while a small amount of amylase reached the circulation by way of the lymph, most of the enzyme was transported by the portal venous route. The importance of the contribution of lymphatics to circulating levels of pancreatic enzymes would be enhanced if the liver removed considerable amounts of pancreatic enzymes presented to it through the portal circulation. The work of Hiatt & Bonorris (1966) has established that the liver can remove considerable amounts of amylase when this enzyme is injected into the circulation of the dog. Singh & Appert (1969), from enzyme measurements in portal and hepatic venous blood in dogs following stimulation of the pancreas with pilocarpine, have also concluded that the liver can take up amylase and lipase.

Howard, Smith & Peters (1949) produced acute pancreatitis in dogs by duct dissection and ligation, and by bile salt injection into the duct. These authors concluded, on the basis of analysis of pancreatic venous, portal venous and aortic blood, that most of the amylase contributing to the marked rise in the level of that enzyme in the blood reached the circulation through the pancreatic venous drainage. Egdahl (1958), however, found in a similar preparation that, in the early phases of the pancreatitis, pancreatic venous blood showed a high concentration of enzymes, but, later in the condition, the level of enzymes in abdominal lymph exceeded the pancreatic venous levels. He also observed that in this pathological process India ink passed through clefts between acinar cells and that duct rupture was not necessarily the route by which pancreatic enzymes reached the interstitial space of the pancreas. Papp, Németh & Horváth (1971) have found that when acute experimental pancreatitis is induced in dogs by injection of bile into the pancreas through a cannula in the duct of Santorini lipase levels in pancreatico-duodenal lymph rise above the values obtained for pancreatic venous blood. This is one of the few studies which has managed to examine the partition of the enzyme between blood and lymph close to the gland, though the relative quantitative importance of the two routes is not clear.

If a significant proportion of the pancreatic enzymes which collect in the peritoneal exudate and in the interstitium of the pancreas in acute pancreatitis is drained by the lymphatic route there are two important consequences which have opposing effects. First, the drainage of toxic substances from the diseased gland may decrease the self-digestive process. Papp, Németh, Feuer & Fodor (1958) have shown in dogs that lymphatic obstruction aggravates the digestive process, and in rats they noted an increase in mortality from trypsin-induced pancreatitis when the thoracic duct was ligated. Dupont & Litvine (1964) have also demonstrated that the severity of the process of pancreatitis following pancreatic duct obstruction in rabbits is greatly increased by lymphatic obstruction. The studies of Sim, Duprez & Anderson (1966) have shown, in dogs with thoracic duct fistulae, that the production of experimental acute pancreatitis results in an initial rise in lymph flow, but this flow falls as red cells appear in the lymph and may finally cease altogether, presumably because of obstruction of draining lymph vessels by aggregates of these cells or by intravascular coagulation. If adequate lymph drainage exerts a beneficial effect on the progress of the disease locally, then this latter sequence of events would aggravate the pathological process. Papp, Varga & Makara (1973) and Papp, Makara & Folly (1975) have shown that a rapid infusion of physiological saline into the superior pancreatico-duodenal artery of dogs results in gross oedema of the pancreas with a substantial rise in the pressure in the pancreatico-duodenal lymphatics, and they consider that peripancreatic lymph nodes may offer a significant obstruction to high rates of pancreatic lymph flow.

A second, but in this case deleterious, result may arise from lymphatic transport of enzymes and toxic agents from the pancreas. These substances may reach the general circulation in high concentration without initial passage through the liver where they might be extracted or inactivated. Lymphatic dissemination of these toxic substances may be important in the production of metastatic necrotic lesions in organs such as the lungs and bone marrow. Brzek & Bartos (1969) have studied the therapeutic effect of prolonged diversion of thoracic duct lymph in patients with acute pancreatitis. These authors found that lymph flow was high in the early stages of the disease (see Sim *et al.*, 1966) and that lymph amylase concentrations were higher than those of blood. Prolonged drainage of lymph appeared to be of considerable benefit in the treatment of the condition. Considerable amounts of pancreatic enzymes appeared to be delivered to the circulation by the lymph, and the favourable effect of lymph diversion on the course of the condition was thought to be due to the removal of lipolytic and proteolytic enzymes, together with digestion products, such as vaso-active peptides and the lysolecithins produced by the action of pancreatic phospholipase on tissue lecithin. These latter substances are powerful surfactants which exert toxic activity by solubilising lipids of cell membranes.

As hypovolaemia is a frequent accompaniment of acute pancreatitis and contributes to the circulatory collapse which often ensues, it is evident that adequate fluid, electrolyte and protein replacement should accompany lymph drainage. Sim *et al.* (1966) suggested that obstruction of pancreatic lymphatic vessels by aggregates of red cells might, by blocking an important route for fluid and protein return to the circulation, contribute to this hypovolaemia.

The possibility that an anatomical connection between gall bladder lymphatics and pancreatic lymph vessels can be responsible for the occurrence of an acute pancreatitis in association with acute biliary tract inflammation has been considered since the beginning of this century. A recent study by Weiner *et al.* (1970) has re-investigated this problem. India ink infusion into the lymph vessels of the gall bladder or pancreas in dogs has demonstrated an anatomical communication between these vessels. In dogs an experimental inflammation of the gall bladder led to a high incidence of pancreatitis and a lipase-induced pancreatitis was frequently accompanied by acute cholecystitis.

In summary, it appears that pancreatic enzymes and their digestion products pass into the general circulation by both venous and lymphatic routes. The drainage area must be considered to be both the necrotic gland and the peritoneal cavity in which an exudate is present. The lymphatic route gains its greatest significance where there is a high rate of lymph flow and where particulate matter, cells and macromolecules, are being transported. The filtering action of the liver acting on digestive enzymes transported in blood may add to the relative importance of the lymphatic route in the transport of enzymes and toxins to the general circulation when compared with the portal venous route. The beneficial therapeutic effect of lymph diversion in acute pancreatitis further supports the idea that there is an important degree of lymphatic transport of these substances in this condition.

Enzymes liberated from gastro-intestinal tissues during shock

Liberation of enzymes from tissues undergoing pathological change occurs in many situations. It is probable that some transport of enzymes from the site is by lymphatics since this is an important route for the transport of large molecules. Haemorrhagic shock offers another example of a condition in which autolysis of gastro-intestinal tissues occurs. In this condition, hepatic parenchymal cells and intestinal epithelium are both very vulnerable. These cells contain large numbers of lysosomes whose degradative enzymes are released during the autolytic process. With the hypothesis that the lymphatics would transport such liberated enzymes, Dumont & Weissmann (1964) examined the transport of the lysosomal enzyme β-glucuronidase in the thoracic duct lymph of dogs during haemorrhagic shock. Their results suggested that β-glucuronidase is transported to the blood via lymphatics, and since the diversion of thoracic duct lymph prevented any rise in serum β-glucuronidase, it was thought likely that in this condition little of the enzyme reaches the circulation by splanchnic veins. This conclusion has been challenged by Barankay, Horpácsy, Nagy & Petri (1969), who found in their dogs that thoracic duct lymph diversion did not prevent a rise in lysosomal enzymes in the blood during haemorrhagic shock. In their study, in which three lysosomal enzymes were measured (acid phosphatase, β-glucuronidase and leucine aminopeptidase) the blood and lymph enzyme levels were followed over a three hour period. It was found that the blood enzyme levels were consistently higher than those of the lymph.

The discrepancy between these two studies may lie in the form of hypotension created in the animals. In the study of Dumont & Weissmann (1964), the animals were bled to death over 57–90 min; no blood pressure record is given. In the Hungarian study, the mean arterial pressure was maintained throughout the three hour period at 35 mmHg.

Alkaline phosphatase

This enzyme, which catalyses hydrolysis of phosphoric esters, is characterised by its high pH optimum (approximately 9.5). It is present in significant amounts in bone, small intestinal mucosa, kidney, liver, placenta and leucocytes. The enzyme consists of a series of isoenzymes which can be separated electrophoretically and these isoenzymes are characteristic of their tissue of origin, so in electrophoretic separation of alkaline phosphatases of blood serum the contributions from the various tissues can be clearly distinguished. Histological studies with light and electron microscope indicate that in mammals alkaline phosphatases are primarily localised to absorptive or secretory surfaces of cells. In the case of the small intestinal mucosa, this is the microvillous membrane (Watanabe & Fishman, 1964). Such a histological localisation suggests a possible role for alkaline phosphatase in active transport processes. It is possible that in the small intestine, this enzyme corresponds to 'calcium-activated ATPase'. Vitamin D given to rachitic chicks produces a two- to three-fold increase in intestinal alkaline phosphatase concentrations, and this parallels a rise in the transport of calcium across in-vitro preparations of intact ileum. Both these effects are inhibited by substances interfering with protein synthesis, such as cycloheximide (Norman, Mircheff, Adams & Spielvogel, 1970). This has been confirmed by Haussler, Nagode & Rasmussen (1970) who showed that vitamin D given to chicks increases the concentrations of intestinal calcium-activated ATPase.

Many studies point to the participation of intestinal alkaline phosphatase in the process of lipid absorption. For example, Madsen & Tuba (1952) demonstrated increased levels of blood and intestinal mucosal alkaline phosphatase in rats placed on a high fat diet. Inglis, Ghosh & Fishman (1968) found that a breakfast containing a large amount of fat induced a transient rise in the concentration of the intestinal isoenzyme in the serum of human subjects. A similar finding was reported by Warnock (1968), and she also found that this rise in serum alkaline phosphatase did not occur after a meal of skim milk or glucose. Bile salt micellar solutions of fatty acid in the duodenum cause an increase in the alkaline phosphatase activity in the lumen of the duodenum in rats, and the enzyme under these conditions appears to migrate from the brush border of the mucosal epithelium to the lymphatics of the mucosa (Bluestein, Malagelada, Linscheer & Fishman, 1970).

It has been known for a long time that during fat absorption alkaline phosphatase concentrations in intestinal lymph rise. It will also be recalled that increases occur in the concentrations of tributyrinase and a non-specific esterase in rat intestinal lymph following fat feeding. Early investigations of the rise in intestinal lymph alkaline phosphatase in rats during fat absorption were made by Flock & Bollman (1948, 1950). In these studies, it was found that lymph levels of the enzyme exceeded plasma levels, and it was concluded that alkaline phosphatase derived from the intestinal mucosa was delivered to the circulation by the lymphatic route, as lymph drainage greatly reduced plasma levels of the enzyme. These authors also noted that biliary diversion reduced the rise in lymph alkaline phosphatase in response to fat feeding and concluded that bile somehow mediated the release of intestinal alkaline phosphatase into lymph. In man, also, alkaline phosphatase activity in thoracic duct lymph rises after feeding (Linder & Blomstrand, 1958). Blomstrand & Werner (1965) studied the concentrations of alkaline phosphatase and triglyceride in human thoracic duct lymph after a single fat meal, finding a close correlation between them. Dietary carbohydrate and protein did not cause an elevation of lymph alkaline phosphatase. Studies with L-phenylalanine, an inhibitor of *intestinal* alkaline phosphatase, indicated that the major portion of lymph alkaline phosphatase is of intestinal origin.

Keiding (1964) reached a similar conclusion from electrophoretic studies of alkaline phosphatase present in human lymph. An alkaline phosphatase moving as a slow β-globulin, a β_2-phosphatase, was found in thoracic duct lymph. A peak in the concentration of this β_2-alkaline phosphatase was observed within four to seven hours after a cream meal, and the concentration of the enzyme in lymph rose in proportion to the increase in lymph lipid (Fig. 5.7). This phosphatase, which was also found in intestinal fluid and, to some extent, bile, was rarely found in peripheral blood and was not demonstrable in

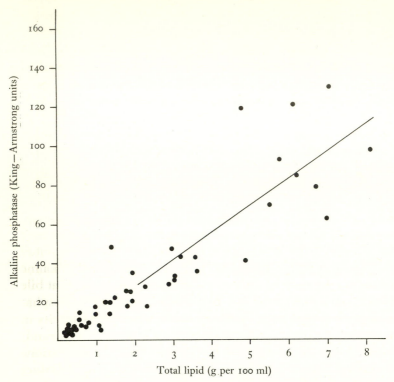

Fig. 5.7. The relationship between alkaline phosphatase activity and lipid concentration in different samples of thoracic duct lymph in nine patients. (Keiding, 1964.)

portal venous blood. This latter observation suggests a specific transport of the enzyme from mucosal cells into intestinal lymph. These studies also suggest that the alkaline phosphatase delivered to the circulation by way of the lymphatics during the absorption of a fat meal in man must be rapidly cleared, possibly in part by excretion in the bile or by transformation in the liver since little of the β_2-phosphatase is found in post-prandial serum.

In rats, the rise in alkaline phosphatase in serum after feeding is due, in part at least, to delivery of intestinal alkaline phosphatase to the circulation. This has been demonstrated by studies with rat intestinal alkaline phosphatase antibody (Saini & Posen, 1969). On the basis of other criteria, such as heat resistance, electrophoretic migration in starch gel and L-phenylalanine sensitivity, it seems clear that in this species a post-prandial elevation of serum alkaline phosphatase is due to delivery of intestinal alkaline phosphatase to the circulation. Intestinal alkaline phosphatase in the rat circulation has a short life since the enzyme injected intravenously disappeared from the circulation of recipient rats within two hours.

The mechanism by which fat absorption brings about a rise in intestinal alkaline phosphatase in lymph has been studied by Glickman, Alpers, Drummey & Isselbacher (1970). These authors, using rats with mesenteric lymph fistulae, demonstrated the close correlation between intestinal lymph alkaline phosphatase and triglyceride concentrations both in the basal state and after feeding bile salt–micellar solutions of oleic acid. After intraduodenal administration of the lipid, the rise was observed within the first hour. The contrast of this short period with the much longer periods of delay before peak lymph alkaline phosphate levels are reached after feeding as reported by other workers, is probably because, in this study, the lipid was infused as free fatty acid into the duodenum as opposed to administration of triglyceride into the stomach. When octanoic acid was given as a micellar solution, the lipid was primarily transported by portal venous blood, and no significant increase in lymph triglyceride transport occurred. A rise in lymph alkaline phosphatase nevertheless occurred, indicating that the process of fat absorption was responsible for the observed effects and not that a lipid molecule was involved in the transport of the enzyme into lymph, as had been suggested by Keiding (1964). The observation by Flock & Bollman (1950), that biliary diversion reduced the rise in lymph alkaline phosphatase in response to a fat meal, was probably due to impaired fat digestion and absorption occasioned by the absence of bile salt. The studies of Glickman *et al.* (1970) also indicated that the rise in lymph alkaline phosphatase after feeding a fat meal is not due to absorption of biliary alkaline phosphatase since the phenomenon occurred in rats with biliary fistulae given micellar solutions of lipids by intraduodenal infusion. Another possible mechanism, a general 'leakiness' of the small intestinal mucosal epithelium during fat absorption, was discounted since no significant changes in lymph levels of 'marker' mucosal enzymes, such as acid phosphatase, leucine amino-peptidase and lactase, were observed while fat absorption was in progress. In this same study, the question of whether preformed or newly synthesised intestinal alkaline phosphatase is liberated from the mucosal epithelium during fat absorption was examined. Using acetoxycycloheximide as an inhibitor of protein synthesis, they observed that duodenal infusion of oleic acid micelles brought about a rise in lymph triglyceride transport with no accompanying rise in alkaline phosphatase. The increase in lymph triglyceride transport in this experiment was of such extent as would ordinarily be accompanied by increased lymph alkaline phosphatase (Fig. 5.8). These results led to the conclusion that new synthesis of alkaline phosphatase occurs during fat absorption, and the release of this enzyme into intestinal lymph is, in some way, connected with the uptake or transport through the mucosal epithelial cell of dietary lipid. While the significance of this is still unclear, these experiments suggest that the release of this enzyme into lymph is not an essential part of the normal processes of fat absorption.

On the other hand, Linscheer, Malagelada & Fishman (1971) obtained results in man which suggest that alkaline phosphatase activity in the small intestine is essential for normal absorption of oleic acid. In these studies, a segment of the small intestine in human subjects was perfused with a micellar solution of labelled oleic acid and the effect of addition of L-phenylalanine and β-glycerophosphate to the perfusion medium was examined. This combination of an inhibitor of intestinal alkaline phosphatase and a substrate for the enzyme produced an 80% inhibition of enzymic activity. Such a combination produced

---

Fig. 5.8. (*a*) Effect of intraduodenal infusion of micellar solutions of oleic and octanoic on lymph alkaline phosphatase and triglycerides (TG) in rats. After a control period of one hour, oleic or octanoic acid micellar solutions (5 ml) were infused over a 30 min period. Solid lines represent mean enzyme and triglyceride values; shaded areas represent range of individual determinations. The number of animals is indicated in each figure. (*b*) Effect of acetoxycycloheximide (ACH) on lymph alkaline phosphatase and triglyceride (TG) after infusion of oleic and octanoic acid micellar solutions. As shown acetoxycycloheximide was given intraperitoneally one hour prior to the lipid infusion. The number of animals is indicated in each figure. (From Glickman *et al.*, 1970.)

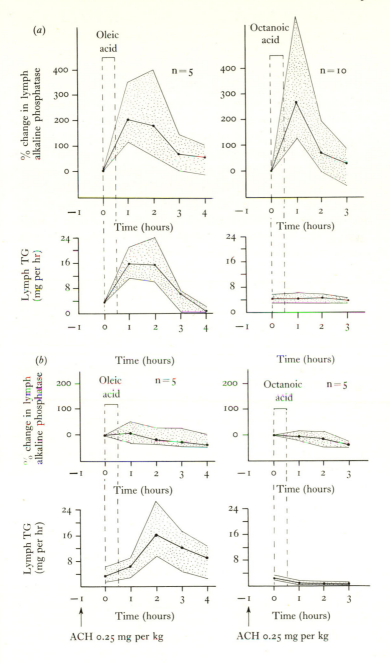

a very marked inhibition of oleic acid uptake. These experiments, unfortunately, did not include control studies of absorption of non-lipid nutrients under similar circumstances. Lam & Mistilis (1973) found that there was a small rise in intestinal lymph triglycerides following administration of intraduodenal octanoic acid in rats and, in agreement with Glickman *et al.* (1970), found a rise in lymph alkaline phosphatase. Oleic acid produced a large increase in lymph triglycerides and alkaline phosphatase. Thus, intraluminal *digestion* of fats is not necessary to bring about the increase in alkaline phosphatase in lymph. Studies of uptake of labelled fatty acids by the mucosa suggested that it was not fat digestion which was responsible for the increase in alkaline phosphatase levels. Thus it seems likely that the increase is related either to triglyceride (or lipoprotein) synthesis within the enterocyte, or else to the passage of lipid (or lipoprotein) through the cell. This passage involves the membranous compartments within the cell and the process of reversed pinocytosis. Alkaline phosphatase was not found to be a structural component of lipoprotein particles in lymph. The process of fat digestion and absorption is also accompanied by the release of diamine oxidase (histaminase) from the intestine. In the rat, olive oil feeding results in a marked rise in the activity of this enzyme in intestinal lymph (Wollin & Jacques, 1974). If the release of this enzyme were viewed as a response to histamine release, that is, a means of limiting the action of histamine, this might indicate the participation of histamine in the changes in intestinal blood and lymph flow during fat assimilation (see Chapters 3 and 7).

Summary

The phenomena of digestion in the alimentary tract appear to involve the participation of the lymphatic system in a rather limited sense. Certain secretagogue hormones and inhibitory hormones can be shown to be present in lymph, and those which are macromolecules or are bound to macromolecules may be transported in part in lymph. The bulk of evidence indicates, however, that the most studied of these, gastrin, does not reach the circulation by lymph in physiologically significant amounts during digestive activity.

Digestive enzymes appear in modest amounts in peripheral serum and probably are partly transported via lymphatics to the circulation. In pathological conditions, notably acute pancreatitis, the oedematous gland and peritoneal exudate contain large amounts of pancreatic enzymes and toxic products of tissue damage. The lymphatic system undoubtedly plays a major part in the systemic absorption of these substances. The fascinating observations on the lymphatic transport of intestinal alkaline phosphatase and diamine oxidase during fat absorption need further study. Clearly, alkaline phosphatase is involved in some way with the process of fat absorption rather than fat digestion, but as yet, the nature of this participation is unclear.

6 TRANSPORT OF ABSORBED WATER AND WATER-SOLUBLE SUBSTANCES

While it is widely acknowledged that the intestinal lymphatics play a specific part in the transport of absorbed lipids and lipid-soluble substances, their importance in carrying water and water-soluble substances absorbed by the intestine is less clear. In general the lymphatic system affords a route of transport for excess tissue fluid and protein and hence might be expected to participate in the disposal of water and solutes accumulating in the interstitium of the intestinal villus during absorption. If it does, the participation of the intestinal lymphatics in the transport of absorbed water and solutes might merely represent the general function of the lymphatic system acting in a special-ised tissue. The endothelial walls of blood capillaries and terminal lymphatics are thought to afford little hindrance to the entry of small solutes and water into these vessels. One might therefore expect that the fast flowing portal venous drainage of the intestine, estimated to be as much as 500 times greater than the flow of intestinal lymph (Bollman *et al.*, 1948; Reininger & Sapirstein, 1957), should account for the transport of major amounts of water and small solutes. Since the lymph-atics act as an overflow system, the amounts of fluid carried by them can vary.

Oral lymphagogues

Many different substances given by mouth increase lymph flow (Yoffey & Courtice, 1970). In rats the lymphagogic effects of olive oil (Tasker, 1951) and corn oil (Borgström & Laurell, 1953) have been observed and are discussed in Chapter 7. Simmonds (1954) found enhanced flows of lymph in the cisterna chyli after intragastric administration of water, 0.9% sodium chloride, 10% serum albumin or 10% glucose to rats (Fig. 6.1). Of these

126

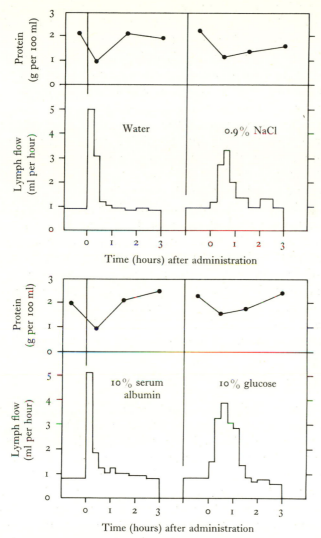

Fig. 6.1. The effects of 5 ml of water, 0.9% sodium chloride, 10% serum albumin and 10% glucose introduced into the stomach in a single dose, on the flow and protein concentration of lymph collected from the thoracic duct of the unanaesthetised rat. (Simmonds, 1954.)

isotonic sodium chloride is a particularly potent lymphagogue and some workers use an intragastric infusion of this solution to maintain good mesenteric lymph flows in the period immediately after creating a lymph fistula. In view of the heterogeneous nature of these various lymphagogues, it is probable that they operate through different mechanisms to increase the volume of lymph. Clearly, the increased flow might be due to the absorption of fluid as would occur following the administration of water, a glucose solution or saline. Gallo-Torres & Miller (1969), however, have proposed that many different oral lymphagogues exert their effects by promoting a sodium-dependent transport of fluid from the intestinal lumen. More work is needed to substantiate this hypothesis.

It is probable that the lymphagogic effect of a triglyceride meal is due to increased filtration of fluid from the intestinal blood vessels as a result of circulatory alterations during digestion and absorption. During absorption substantially increased amounts of protein are transported per unit time through the thoracic duct (Borgström & Laurell, 1953). This could be explained by increased filtration of plasma protein during such circulatory changes together with the addition of newly synthesised lipoproteins of mucosal origin (see Chapter 7).

Absorption of water

In several species, oral intake or intragastric administration of water has been shown to increase lymph flow. This has been demonstrated in anaesthetised dogs (Watkins & Fulton, 1938), cats (Korner, Morris & Courtice, 1954) and in a conscious human subject (Crandall, Barker & Graham, 1943). In the conscious rat with a thoracic duct fistula, Simmonds (1954) found that intragastric administration of water over 10 min caused a rapid rise in lymph flow with an accompanying dilution of its protein (Fig. 6.1). Similar effects were found in anaesthetised cats (Korner *et al.*, 1954). These results suggest that some water might be transported from the site of absorption by intestinal lymph vessels. Several workers have attempted to evaluate the relative importance of the portal venous and lymphatic routes in the transport of fluid.

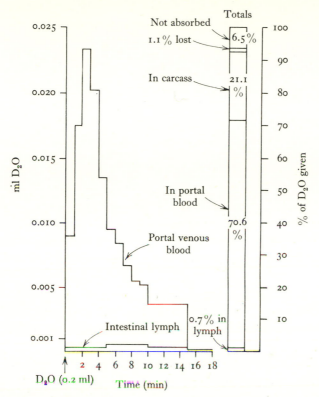

Fig. 6.2. Total amount of D_2O in the portal blood and intestinal lymph each minute after intragastric administration of 0.2 ml D_2O. The column at the right shows the distribution of the total D_2O for the 18-minute period. (Benson *et al.*, 1956.)

In a study of water absorption in unanaesthetised rats with mesenteric lymph fistulae, small doses of heavy water (D_2O) were given and it appeared that about 1% of this water was transported by the lymphatic system (Benson *et al.*, 1956). Fig. 6.2, taken from this study, shows the amounts of D_2O in intestinal lymph and portal venous blood following intragastric administration of 0.2 ml D_2O. Fig. 6.3, also from this investigation, shows the concentrations of D_2O in arterial plasma and intestinal lymph during absorption of D_2O from the gastrointestinal tract. These authors considered that the lymph water was derived from arterial plasma. Noyan (1964), using tritiated water (T_2O), found that the lymphatic route accounted for

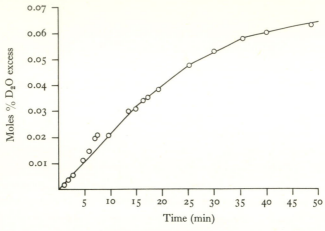

Fig. 6.3. Concentration of D_2O in the femoral arterial plasma, O, and intestinal lymph, ——, during absorption from the stomach and intestine. (Benson *et al.*, 1956.)

about 3% of the transport of absorbed water. This was based on the recovery of T_2O from the mesenteric lymph and body tissues after perfusion of the small intestine of the anaesthetised rat with Krebs bicarbonate buffer solution containing T_2O.

Both of these studies attempt to assess a value for that proportion of water absorbed by the intestine which is carried by intestinal lymph vessels. In view of the function of lymphatics as an overflow system, such experiments can only define the proportion carried under certain specified circumstances, such as those which obtained in these studies. For example, in both cases the absorption of labelled water was studied in a situation in which fluid was being absorbed from an isotonic solution, since in the experiments of Benson *et al.* the flow of intestinal lymph was maintained by intragastric or intraduodenal infusion of solutions of 0.9% sodium chloride or 5% dextrose, while Noyan used Krebs buffer to perfuse the small intestine. As considerable differences exist between flow and composition of lymph when water, isotonic saline or isotonic dextrose are ingested, their results can only be taken to apply to these particular conditions of fluid absorption. Furthermore, the use of isotopic water to quantify the part played by lymphatics in water transport is misleading since the free diffusion of water between various fluid compartments results in rapid equilibration of the isotope with the body water.

In unanaesthetised rats with lymphatic fistulae, water ingestion results in a rapid rise in flow of dilute lymph within 1 min in the cisterna chyli or main mesenteric lymph vessel (Barrowman & Roberts, 1967) (Fig. 6.4). It can also be seen in this Figure that the initial brisk rise in lymph flow results in a flushing out of concentrated 'dead space' lymph with a consequent brief rise in protein output from the fistula. When tritiated water is given, a 'spike' of isotope appears in the intestinal lymph and thereafter the concentration of isotope returns to a level consistent with its distribution throughout the body water (Fig. 6.5). The increase in water flowing in the intestinal lymph vessels after oral intake of 5 ml amounts to approximately 14% of the ingested dose. These results are interpreted as evidence that ingestion of water results in a transient bulk flow of hypotonic fluid into the mucosa of the small intestine and thence to lymph and blood vessels. The absolute amount transported in the lymph in such a situation would be related to the amount of hypotonic fluid presented to the mucosa rather than a fixed proportion of the ingested amount. It is probable that a large drink of hypotonic fluid in man results in a similar substantial flow of intestinal lymph. It is likely that water flows across the mucosa from a hypotonic solution in the upper small intestine, while electrolytes pass in the opposite direction so rendering the intestinal contents isotonic with plasma in a short time. Such effects have been demonstrated in the small intestine of the rat by Follansbee (1945). In this study, rats which were deprived of food, but not fluids, for twenty-four hours were given, by intragastric tube, volumes of water amounting to 2% of their body weight. Ten minutes after administration, the fluid in the small intestine was approaching isotonicity with plasma, although gastric contents were still considerably hypotonic. When larger amounts of water were given (3% of the body weight), the duodenal contents at ten minutes had a tonicity approximately two thirds that of plasma, while jejunal contents were only a little less than isotonic with plasma. These results, together with data from Code *et al.* (1960) obtained from dogs, suggest

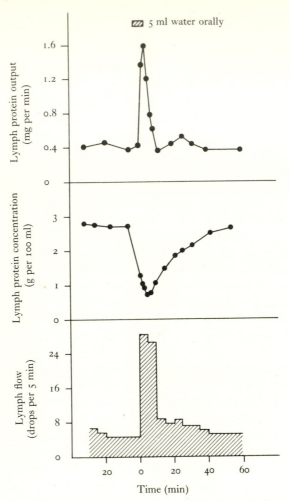

Fig. 6.4. Lymph flow, protein concentration and protein output from a cisterna chyli fistula in a rat after ingestion of 5 ml of water. (Barrowman & Roberts, 1967.)

that the duodenum acts in an equilibrating capacity, tending to produce isotonicity of ileal contents by maintaining large fluxes of water or solute in either direction across the mucosa.

Intragastric infusion of isotonic sodium chloride solution also causes a rise in flow in the intestinal lymph vessels (Kim & Bollman, 1954). However, the rise in flow after a single drink of saline in rats was shown to be slower than that observed after

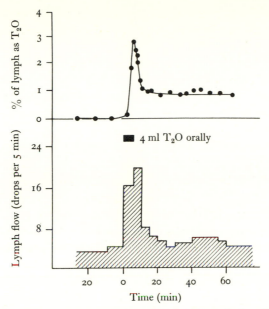

Fig. 6.5. Lymph flow and concentration of T_2O in the main intestinal lymph vessel of a rat following the ingestion of 4 ml of T_2O. (Barrowman & Roberts, 1967.)

water was administered, and the enhancement of flow was more sustained (Barrowman & Roberts, 1967). Protein concentration in this lymph fell as flow rate rose. The absorption of such a solution may involve active solute absorption as an isotonic solution, and since this water uptake would be a slower process than the rapid inflow of hypotonic fluid, the major amount of fluid would probably be taken up by the portal venous blood. The increased flow of lymph in this case might be derived partly from the blood as a result of increased filtration after expansion of the plasma volume, and partly from a rise in mucosal interstitial fluid during the absorption of the saline solution. A single intravenous injection of 5 ml of 0.9 % saline – a volume comparable with those drunk by rats in the experiments of Barrowman & Roberts (1967) – results in a rapid increase in lymph flow in the cisterna chyli of the rat. The protein concentration of this lymph is below control levels. It is likely that this effect reflects rapid equilibration of the saline solution throughout the extracellular compartment. Approximately

133

20% of such an intravenous load is recovered in cisterna chyli lymph in 45 min after injection (Barrowman & Roberts, unpublished observations).

Code & Pickard (1973) have tried to determine the relative importance of lymph versus portal venous blood in fluid transport by infusing full or half-strength Tyrode's solution into the duodenum of three groups of rats: some with intact lymphatics, some with a cannula in the major intestinal lymph vessel and some in which this vessel had been ligated. Table 6.1 shows the flow of urine in these animals during a five-hour period of infusion. From these results it was concluded that net gain of water to the blood was via the lymphatic system and that only when an osmotic gradient arises between blood in intestinal capillaries and the intestinal lumen does the intestinal blood carry the major part of the absorbed fluid. While this interpretation fits the data it is important to recognise that intestinal lymphatic ligation is likely to cause a serious degree of oedema of the lamina propria of the mucosa and possibly impede further absorption of isotonic fluid, whereas hypotonic fluid might in part be able to enter the blood by reducing the osmotic pressure of the interstitial fluid of the mucosa. It should also be remembered that an intestinal lymph fistula drains lymph which has resulted from an expanded plasma volume as well as fluid absorbed directly into lymph.

Lee (1961, 1969a) has approached the question of lymphatic transport of absorbed fluid by studying in-vitro preparations of rat jejunum. Fig. 6.6 shows a preparation of a segment of rat small intestine in which blood and lymph vessels have been cannulated. This preparation was perfused through the gut lumen with a bicarbonate-buffered physiological salt solution containing 0.5% glucose aerated with 95% O_2–5% CO_2. At a lavage pressure of 10 mmHg, lymphatic flow accounted for 85% of the water transport across the mucosa and venous flow for only 11%. When the vein and the lymphatic were occluded, the mean lymphatic pressure was three times higher than the simultaneous venous pressure. Lymph flow decreased when mucosal fluid osmolarity was increased but lymph was always isosmotic with mucosal fluid whether hypo- or hypertonic. Further evidence of the considerable importance of the intes-

TABLE 6.1. *Volumes of urine and lymph collected during a five-hour period of infusion of saline solutions into the duodenum.* (Code & Pickard, 1973.)

Infusion Rate (ml per hr)	Solution	Lymphatics intact. Urine (ml)	Lymphatics ligated. Urine (ml)	Lymphatics cannulated	
				Urine (ml)	Lymph (ml)
8.5	Tyrode full-strength	26.7±2.1	4.6±0.8	5.1±1.7	15.3±1.1
	Tyrode half-strength	28.5±1.4	18.3±2.3	17.9±0.8	12.5±2.3
4.2	Tyrode full-strength	15.2±0.5	0.8±0.2	0.8±0.1	7.4±0.7
	Tyrode half-strength	15.9±0.3	7.9±0.8	3.4±0.4	7.9±0.7

Values are means±s.e.; four rats in each group.

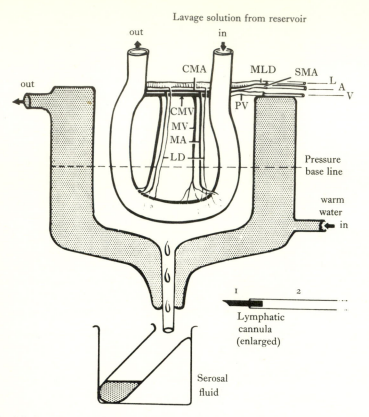

Fig. 6.6. Arrangement of intestinal segment in warm chamber for pressure measurement and collection of various fluids. Abbreviations: A, arterial cannula; CMA, common mesenteric artery; CMV, common mesenteric vein; L, lymphatic duct cannula; LD, lacteal duct; MA, mesenteric artery; MLD, main lymphatic duct; MV, mesenteric vein; PV, portal vein; SMA, superior mesenteric artery; V, venous cannula. 1, 23 gauge needle point; 2, polyethylene tubing. (Lee, 1961.)

tinal lymphatic system in water transport has been obtained using an isolated rat intestine preparation which was essentially similar to that described by Fisher & Parsons (1949). In this study (Lee, 1963) water absorption from a rat jejunal preparation, perfused with bicarbonate-buffered physiological saline, was measured as fluid transported to the serosal surface of the segment; this fluid transport was reduced when the mesentery was absent or where, with an otherwise intact mesentery, the lymphatics were sectioned close to the gut wall.

Blood vessel section did not have this effect. Since both *in vivo* and *in vitro* the mesenteric lymphatics were found to have rhythmic (ten per minute) contractions, it was proposed that the water absorption might be facilitated in these in-vitro preparations by the propulsive activity of the lymphatics, or, alternatively, that spasm of cut lymph vessels would seriously impede water transport. There seems to be no doubt that in these in-vitro preparations lymphatics play a major role in fluid transport. It appears that intestinal motility, lymphatic contractility and the distension pressure of the gut all contribute to the pressure developed within mesenteric lymphatics and to the flow of lymph (Lee, 1965). Drugs such as adrenaline and sodium pentobarbital, which inhibit intestinal motility, were found to decrease lymphatic pressures and flows, and conversely serotonin and physostigmine increased lymph pressures and flows. Adrenaline has a complex action on the system since it reduces lymph pressure and flow by reducing intestinal motility, but subsequently causes a rise in pressure by a direct stimulation of lymphatic contractility (Fig. 6.7). In intestinal segments without motility, the distension pressure of the segment was found to determine lymphatic pressure, probably by

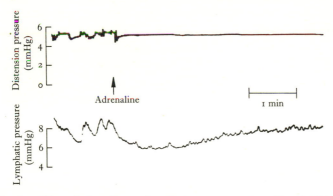

Fig. 6.7. Effect of adrenaline on lymphatic pressure and motility. Addition of adrenaline (0.1 mg per 100 ml) to serosal bathing fluid of an in-vitro preparation of rat intestine abolished motility with the disappearance of distension pressure waves and caused a fall of lymphatic pressure. About a minute later there was a gradual increase of lymphatic pressure with the appearance of small pressure waves. In the control period, the large lymphatic pressure waves were associated with distension pressure waves, indicating the close relationship between lymphatic pressure and motility. (Modified from Lee, 1965.)

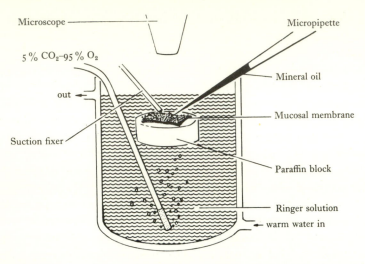

Fig. 6.8. Preparation for cannulation of the central lymphatics of dog jejunal villi. (Lee, 1969*b*.)

increasing the surface area of the mucosa and by improving its oxygenation.

Using micropipettes, Lee (1969*b*) has succeeded in collecting fluid from the central lacteals of intestinal villi in dog jejunal mucosa. The preparation is illustrated in Fig. 6.8. A small piece of mucosal membrane was bathed in bicarbonate-buffered Ringer solution containing glucose and the preparation was exposed to $95\% \ O_2$–$5\% \ CO_2$ mixture. A mean water transport rate of 134 ± 17 nl ($\pm$Standard Error of Mean) per hour per villus was obtained from forty-three measurements. These figures should be regarded as no more than approximate since the technical difficulty of this study is considerable. The preparation is clearly an artificial one since no mucosal blood flow occurs, and the flow rates and composition of lacteal fluid quoted can only be considered to apply to this abnormal situation. As in other in-vitro preparations, the water transport was related to the osmotic pressure of the bathing solution and decreased with increasing osmolarity. It was also noted that water transport depended on the presence of glucose in the bathing solution and it was inhibited by 2,4-dinitrophenol. Occlusion of the tip of the villus prevented water transport, suggesting that this may be the site of water absorption. It is of interest that in villi which were tonically contracted no water transport occurred; this has again been observed in a subsequent study by Lee (1971). The entry of fluid into the central lacteal probably requires relaxation of the villus, while phasic contractions of the villus help to promote the transport of lymph to the deeper lymphatics of the intestinal wall. Lee & Silverberg (1972) have also used this in-vitro preparation to demonstrate reduced fluid transport into the small intestinal mucosa when it is exposed to cholera toxin. In this situation villus lymph pressure is markedly reduced.

These in-vitro preparations, while demonstrating that the lymph vessels can transport absorbed fluid to a considerable extent, do not allow assessment of the proportions of fluid passing by blood and lymph from the intestine *in vivo*. Since a quantitative estimate of lymph and portal blood transport of absorbed fluid demands that the blood supply to the intestine is intact, Lee & Duncan (1968) have described a method for vascular perfusion of an isolated segment of rat upper jejunum with heparinised homologous blood (Fig. 6.9). In this preparation, it is possible to measure the transport of fluid into both venous and lymphatic systems. With bicarbonate-Ringer solution in the intestine, net water movement occurred into both blood and lymph channels, and the volume so recovered accounted for the amounts disappearing from the intestinal lumen. At low venous pressure (0–5 mmHg), lymph flow accounted for 10–25% of net water transport, but when venous pressure was raised above 20 mmHg, net water transport was entirely via the lymphatic vessel (Table 6.2). This in-vitro system represents a much closer approach to the situation *in vivo* than earlier preparations (Lee, 1961), and the values obtained at low venous pressures (10–25% of net water transport) are comparable with those obtained *in vivo* in rats (11–14% of net fluid transport) after ingestion of water or 0.9% sodium chloride (Barrowman & Roberts, 1967).

As regards fluid transport by intestinal lymph vessels during absorption of liquid test meals, one must conclude that the lymphatics fulfil an overflow role during the phase of maximum fluid transport. Their importance is enhanced by hypertension

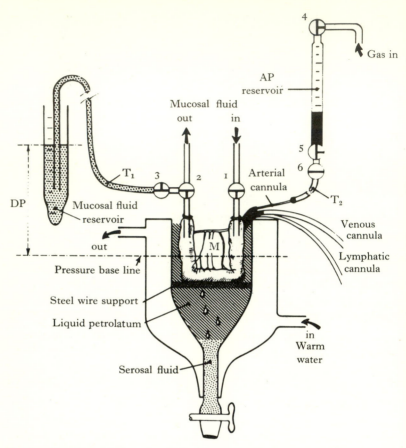

Fig. 6.9. Vascular perfusion of an isolated segment of rat jejunum. M, mesentery; AP reservoir, arterial perfusate reservoir; DP, intraluminal distension pressure; 1–6, stopcocks; T_1, T_2, connecting tubes. (Lee & Duncan, 1968.)

in the portal venous system and while enhanced lymph transport of water is occurring any small water-soluble substances would be expected to be swept into lymph vessels. In obstructive disease of intestinal lymph vessels high rates of water transport might produce an exacerbation of stromal oedema of the villi which are already subject to a chronic lymphoedema due to accumulation of osmotically effective macromolecules such as protein.

The concept of compartments within the interstitial space

TABLE 6.2. *Comparison of various flow rates and water transport rates. (Lee & Duncan, 1968.)*

(Intestine perfused with bicarbonate-Ringer solution for thirty minutes; arterial pressure, 70 mmHg; venous pressure, as shown in table. Values are means ±SEM; n is in parentheses in first column.)

| Venous pressure (mmHg) | Flow rates (μl per cm per hour) | | | | | Water transport rate[a] (μl per cm per hour) | % of water transport rate[a] | |
	Arterial inflow (A)	Venous outflow (V)	(V − A)	Lymph flow	(V − A) + lymph flow		Lymphatic system	Venous system
0 (4)	1782 ± 326	1889 ± 333	107 ± 14	13 ± 0	120 ± 14	120 ± 14	10 ± 1	90 ± 1
10 (5)	1982 ± 250	2075 ± 250	93 ± 18	27 ± 6	120 ± 6	133 ± 42	22 ± 5	78 ± 5
20 (6)	1503 ± 109	1503 ± 114	0 ± 16[b]	53 ± 5	53 ± 11	67 ± 5	100	0

[a] Sum of (V − A) and lymph flow was taken as water transport rate.
[b] In two cases, (V − A) was negative.

of the intestinal mucosa is useful in considering the transport of fluid from the mucosa. Yablonski & Lifson (1976), from studies of intestinal secretion in dogs in response to elevated venous pressure, have proposed that a juxta-capillary compartment exists and the protein concentration of fluid in this compartment would be reduced by incoming water. Changes in hydrostatic and osmotic pressure in this compartment would drive the fluid into the capillaries. Under conditions of high inflow of water into the mucosa the disturbance of interstitial fluid would extend beyond this compartment to the interstitial fluid of the remainder of the lamina propria of the mucosa and under these circumstances fluid would also be driven into the central lacteal.

Sources of protein in intestinal lymph

Macromolecules are taken up from tissue spaces by lymphatic vessels, and such molecules appearing in the interstitial space of the intestinal mucosa will pass into intestinal lymph (Fig. 6.10). Such substances include dietary proteins absorbed intact in the neonate and lipoproteins synthesised in mucosal epithelium. Most lymph protein is derived from the plasma by capillary filtration. Additional proteins, notably immunoglobulins, are added to lymph on passage through lymph nodes; when lymphocytes migrate from post-capillary venules into lymph it is possible that plasma proteins are added to lymph (Quin & Lascelles, 1975). In addition to protein filtered from the blood, another source of protein in intestinal lymph is the huge mass of plasma cells of the lamina propria of the intestinal mucosa. Vaerman & Heremans (1970) have shown that, in the dog, mesenteric lymph contains considerable amounts of immunoglobulin A (IgA) arising from these cells. This immunoglobulin, a large molecule existing in dog serum as a dimer of MW ~ 310000, is the predominant immunoglobulin in intestinal luminal fluid. The mesenteric lymph: systemic blood serum ratio for IgA was found to be much greater than corresponding ratios for the other immunoglobulins, and it is clear that locally synthesised IgA reaches the circulation to a large extent via mesenteric lymph. These

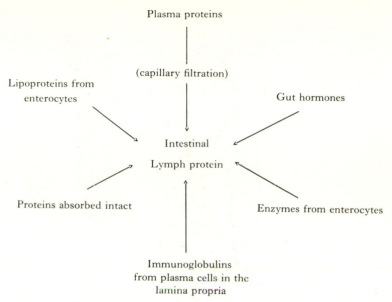

Fig. 6.10. Sources of protein in intestinal lymph.

authors calculated that between 60 and 90% of IgA of canine mesenteric lymph arises directly from mucosal plasma cells. Fig. 6.11 is a proposed scheme of the fate of IgA synthesised in these cells. These same authors have also examined the IgA of the mesenteric lymph of guinea pigs and rats (Vaerman, André, Bazin & Heremans, 1973). The mean concentration of IgA in mesenteric lymph in guinea pigs was 4.5 times that in serum. In rats the concentration in the lymph was 13.4 times that of serum. For dogs the ratio is about 2.4. For proteins other than IgA, lymph:serum ratios are less than 1 (see Table 6.3) and these ratios are inversely related to the molecular weights of the proteins, suggesting that their presence in mesenteric lymph results from filtration of plasma protein from the blood stream. Immunoglobulins labelled with radioactive iodine injected intravenously in sheep, have been measured in intestinal lymph to determine the origin of immunoglobulins of this lymph (Quin, Husband & Lascelles, 1975). Low specific activity of a particular immunoglobulin in the lymph indicates that little of that immunoglobulin has been derived from the plasma

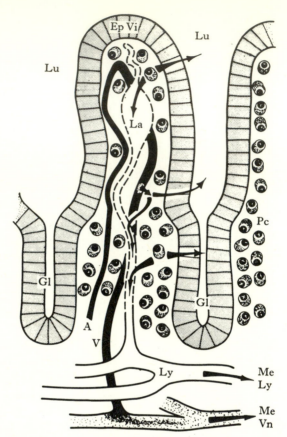

Fig. 6.11. Idealised section of intestinal mucosa showing anatomical origin of mesenteric lymph. Arrows indicate possible directions of transfer of IgA originating either from the blood or the local plasma cells. Lu, intestinal lumen; Gl, glandular crypts; Ep Vi, epithelium of the villi; Pc, plasma cells of the lamina propria; A, arterial capillary; V, venous capillary; La, central lymphatic vessel of the villus, freely permeable to proteins from interstitial fluid; Ly, lymphatic vessel; Me Ly, mesenteric lymphatic; Me Vn, mesenteric vein. (From Vaerman & Heremans, 1970.)

by filtration but that most has been synthesised locally. It appears by this technique that about 90% of IgA in intestinal lymph is locally synthesised and about 25% of the IgG_1 and IgG_2 is also derived from the intestine. The IgG_1 is probably synthesised in plasma cells of the lamina propria of the intestine and to some degree is derived from mesenteric lymph nodes, which are the probable source of most of the IgG_2 of intestinal

TABLE 6.3. *Mesenteric lymph:serum concentration ratios for different proteins in guinea pigs and rats.* (Vaerman *et al.*, 1973.)

Species	Albumin	Transferrin	IgG$_1$	IgG$_2$	IgA	α-Macro globulin	IgM
Guinea pig ($n = 9$)	0.613 (±0.081)	0.534 (±0.078)	0.487 (±0.086)	0.451 (±0.069)	4.453 (±2.303)	0.408 (±0.091)	0.281 (±0.099)
Rat ($n = 10$)	0.688 (±0.131)	0.587 (±0.117)		0.321 (±0.038)	13.375 (±5.524)	0.267 (±0.046)	0.387 (±0.070)

Ratios listed in this table are means of concentration ratios of lymph:arterial and lymph:venous serum±standard deviation.

lymph. Thus it appears that mesenteric lymphatics are a route of primary transport of newly synthesised IgA to the blood and it has been calculated that 90–97% of IgA in mesenteric lymph is synthesised in intestinal immunocytes.

Absorption of proteins and amino acids

Dietary proteins are extensively hydrolysed in the small intestine and absorbed as amino acids or possibly as small peptides. The intestine of adult mammals is virtually impermeable to macromolecules, such as proteins. However, many studies suggest that *small* amounts of ingested proteins may be absorbed intact and reach the general circulation. This has been most clearly demonstrated in the case of proteins which either provoke immune responses or confer passive immunisation in the host after absorption. Bullen & Batty (1957) showed that when concentrated diphtheria antitoxin was dripped into the duodenum of normal sheep small but constant amounts were absorbed and appeared in the blood. It is probable that the lymphatics act as a major route of transport for such macromolecules to the systemic circulation.

In 1936, Alexander, Shirley & Allen, first proposed that proteins absorbed intact from the gastro-intestinal tract are transported thence by lymph. This conclusion was based on antigenic studies of thoracic duct lymph and portal venous blood after feeding egg white proteins to dogs by stomach tube. There is clear evidence that the toxins of *Clostridium botulinum* are transported in the lymph from their site of intestinal absorption (May & Whaler, 1958). Several types of botulinal toxin exist and molecular weights estimated for these various toxins range between 9000 and 900000. Sub-units of these macromolecules have been obtained by proteolytic digestion and the smallest toxic sub-unit prepared by peptic digestion is approximately 3800 – still a comparatively large molecule (Lamanna & Carr, 1967). May & Whaler (1958) studied the absorption from the alimentary tract of *C. botulinum* toxin type A in rats, rabbits and mice. The upper small intestine appeared to be the site of maximal absorption. During its absorption, the toxin appeared in high titres in the thoracic duct lymph of rats and the levels were often greater than those of peripheral blood. The type A toxin has an estimated molecular weight of 900000. In rabbits more than half of the absorbed toxin was recovered in thoracic duct lymph. In rats cannulation of the main intestinal lymph duct with continuous lymph diversion afforded a considerable measure of protection to animals given toxin by mouth. These results point to the intestinal lymphatics as the major transport route for the toxin in these species studied. Such conclusions may apply to a large range of macro-molecular toxins. For example, in a study of the effect of *E. coli* endotoxin in rats, Gans & Matsumoto (1974) showed that when the endotoxin was introduced into a Thiry–Vella fistula a small amount, sufficient to kill a proportion of the rats, was absorbed. The amount absorbed was much less than the amount which the rat liver can clear when the toxin is delivered exclusively by the portal vein, suggesting that a proportion of the toxin bypasses the liver, possibly via the lymphatic route.

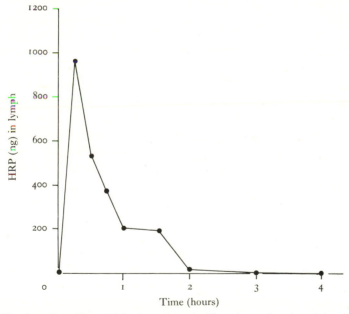

Fig. 6.12. Peroxidase activity in rat intestinal lymph as a function of time following infusion of horseradish peroxidase (HRP) (28 mg per kg) through jejunostomy cannula. This experiment shows the early rise pattern. Each point denotes the amount of enzyme in the total lymph collected during the time interval. (Warshaw *et al.*, 1971.)

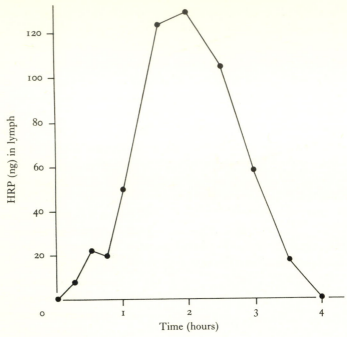

Fig. 6.13. A similar experiment to that shown in Fig. 6.12 showing a typical late response curve. The dose of horseradish peroxidase in this experiment was 32 mg per kg. (Warshaw *et al.*, 1971.)

Such evidence for the importance of lymph as a route of toxin transport can only be considered as very indirect but in view of the capacity of lymphatics to transport toxins from tissues (see Chapter 3) this hypothesis seems reasonable.

Horseradish peroxidase is a convenient substance for studying the phenomenon of macromolecular absorption and transport. It is a protein of MW ~ 40000 having a molecular diameter of about 5 nm, which can be assayed with high sensitivity and demonstrated cytochemically both with light and electron microscopy. Warshaw, Walker, Cornell & Isselbacher (1971) infused the enzyme into the jejunum of rats and were able to demonstrate its presence in intestinal lymph and portal venous blood. Fig. 6.12 and 6.13 show the time course of the appearance of horseradish peroxidase in intestinal lymph. In some animals, an early rise (Fig. 6.12) occurred, while in others (Fig. 6.13) a later and more prolonged transport was observed. It was not possible in this study to define the proportions of absorbed enzyme transported by either route. Cytochemical evidence suggested that most of the absorbed enzyme was transported across the mucosal epithelium by pinocytosis rather than by passive diffusion across effete epithelial cells at the tips of the villi, or passage through disrupted intercellular junctions in the same region.

Studies with ^{131}I-labelled elastase (MW 24000) suggest that the lymphatic route may be important in the transport of small amounts of this enzyme which are absorbed by the intestine. Katayama & Fujita (1972*b*) showed that for seven hours after the labelled enzyme was delivered into the upper jejunum of rats the concentration of immunoprecipitable radioactivity in lymph was ten times higher than that in serum, suggesting direct absorption via lymphatics. Only small amounts of the enzyme were absorbed; totally 0.149% and 0.053% of a 1 mg and 5 mg dose respectively. Of these amounts, 64% was calculated to enter the portal vein and 36% the lymphatic system. Approximately 2% of a dose of ^{3}H-labelled bovine serum albumin introduced into the duodenum of rats with mesenteric lymph fistulae was found to be transmitted to lymph and blood (Warshaw, Walker & Isselbacher, 1974). Over a period of three hours 0.76% and 1.11% of a dose of ^{3}H-labelled bovine serum albumin given intraduodenally to rats were recovered in mesenteric lymph and portal venous blood respectively (Warshaw & Walker, 1974). The concentrations of the labelled protein in lymph did not parallel those in portal blood (Fig. 6.14) in that portal venous concentrations rose steadily while lymph concentrations rose rapidly above venous blood levels and fell to very low levels after two hours. To ensure that isotope measurement represented intact ^{3}H-labelled albumin, all lymph and blood samples were dialysed against large volumes of saline to remove labelled protein fragment and free isotope, while gel filtration and radio-immunoassay against monospecific rabbit anti-bovine serum albumin showed that the radioactive material measured in lymph and blood represented intact bovine serum albumin. This study clearly indicates the existence of the two independent routes of transport of absorbed protein. In young animals, small amounts of

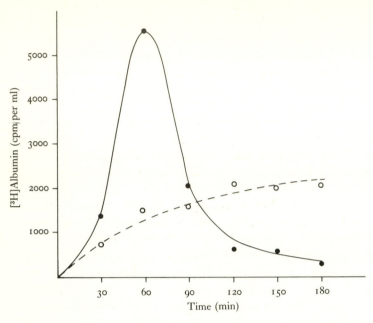

Fig. 6.14. Specific activities of dialysed intestinal lymph, ●———●; and portal plasma, O – – – O at intervals after infusion of [³H]bovine serum albumin into the duodenal lumen. (Warshaw & Walker, 1974.)

intact insulin are absorbed from the intestine but in experiments with newborn calves most of a dose of insulin absorbed from the duodenum entered the bloodstream directly (Pierce, Risdall & Shaw, 1964).

Proteins such as lipoproteins, which are newly synthesised in intestinal mucosa are transported in intestinal lymph (see Chapter 7). Jacobs & Largis (1969a) have studied the absorption of amino acids into lymph showing that, when labelled amino acids together with cream are fed by stomach tube to rats, some of the radioactivity appears rapidly in the free amino acids of lymph and some is incorporated into newly synthesised lymph proteins, probably lipoproteins (Fig. 6.15).

In Fig. 6.16, the specific activity of radioactive lymph protein after feeding [¹⁴C]leucine is shown; this radioactivity was shown on hydrolysis of the protein to be exclusively present in the specific amino acid fed. The concentration of lymph

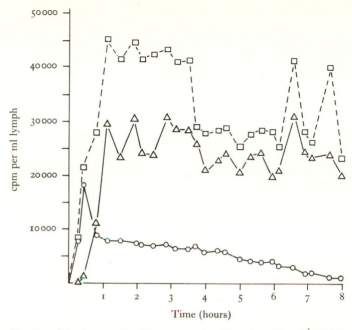

Fig. 6.15. Distribution of radioactivity in lymph after feeding [^{14}C]leucine in cream at time zero. Fractions of lymph (0.5 ml) were collected over an eight-hour period. Total lymph radioactivity, □; protein activity, △; and non-protein activity, ○, the amino acid fraction, are indicated. (Jacobs & Largis, 1969a.)

amino acids was found to be similar to that of plasma. These experiments indicate that amino acids absorbed in a mixed meal enter the pool of amino acid in the mucosal cells and are readily incorporated into protein travelling in intestinal lymph. There is no evidence for preferential transport of amino acids from the intestine by the lymphatics as opposed to the portal vein. Peters & MacMahon (1970), studying the absorption of glycine and glycine oligopeptides in the rat, found levels of glycine and diglycine in mesenteric lymph closely similar to those in portal venous plasma when the amino acid and peptide were infused into the duodenum of rats. It would therefore seem likely that the free amino acids present in intestinal lymph are largely derived from blood and from the diet via the mucosal cell pool.

Jacobs & Largis (1969b) investigated the effects of inhibitors

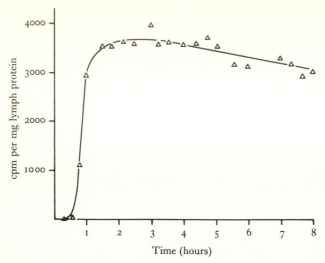

Fig. 6.16. Specific activity of radioactive lymph protein after feeding [¹⁴C]leucine in cream (see Fig. 6.15 for radioactivity distribution in the lymph). (Jacobs & Largis, 1969a.)

of protein synthesis, puromycin and acetoxycycloheximide, on the content of amino acids and newly synthesised protein in mesenteric lymph after rats had been fed meals similar to those described above. Both drugs reduced the appearance of new protein in the lymph, while the concentration of free amino acid in the lymph rose; cycloheximide had a longer action in this respect than puromycin. Isselbacher (1966) has shown that both cycloheximide and puromycin interfere with the transport of lipoprotein and triglyceride from the mucosal cell but as there is some doubt about the mechanisms by which these inhibitors influence fat absorption (see Chapter 7), these results must be interpreted with caution. Fig. 6.17 (Jacobs & Largis, 1969b) is a diagrammatic representation of a proposed scheme of the relationship between fat absorption and protein synthesis in the mucosal epithelial cell.

Absorption of proteins in the neonate
The newborn of various species are able to absorb intact protein from the small intestine for a short period after birth. An important aspect of this phenomenon is the intestinal

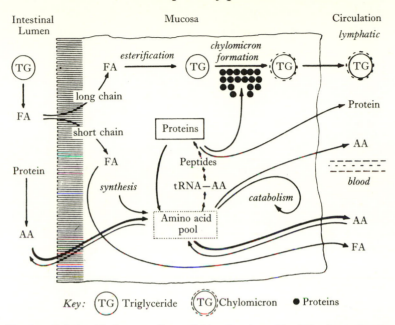

Fig. 6.17. Relationship between dietary lipid and amino acids of the diet and amino acid pools of the mucosal cell and circulation. Abbreviations: FA, fatty acids; AA, amino acids. This simplified diagram does not take into account the role played by phospholipid, bile salts or carbohydrate metabolism. It points out where the puromycin may act – at the tRNA–AA stage – and cycloheximide – at the peptide stage – in the synthesis of protein required for lipid transport into lymph. (Modified from Jacobs & Largis, 1969*b*.)

absorption of immunoglobulins of colostrum (Brambell, 1958, 1970). This mechanism of transfer of immunity is well recognised in ruminants and in the horse and pig. Transport of these macromolecules across the intestinal epithelium is probably by pinocytosis. The importance of lymphatics in the transport of these proteins has been demonstrated in anaesthetised newborn calves (Comline, Roberts & Titchen, 1951*a*). Globulins were measured in lymph, using specific antibody markers which labelled the globulins of colostrum introduced into the small intestine. Unchanged globulins of colostrum did not enter the portal circulation in any appreciable amounts but were found to be carried in the lymph. Simpson-Morgan & Smeaton (1972), using labelled albumin and γ-globulin, have shown that uptake of these proteins from the small intestine

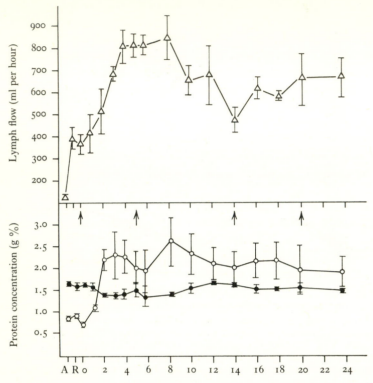

Fig. 6.18. Lymph flow and concentration of albumin and globulin in thoracic duct lymph immediately before and during the twenty-four hour period after first feeding. The points plotted are means±standard errors for the results of three calves. Arrows indicate the times at which the calves were fed colostrum. Symbols: △——△, flow in ml per hour; ○——○, globulin concentration (g%); ●——●, albumin concentration (g%); A, during anaesthesia; R, during the recovery period. (Shannon & Lascelles, 1968*b*.)

into intestinal lymph can occur in the foetal lamb several days before the end of gestation. Shannon & Lascelles (1968*b*) have also studied this phenomenon in newborn calves with thoracic duct–venous shunts finding that during the eight hours after the first feed of colostrum there was a rise in lymph flow and lymph globulin concentration (Fig. 6.18). The globulin was shown by gel filtration and ion-exchange chromatography to be γ_1-globulin, absorbed from the colostrum in which it is present in high concentration. Their results suggested that by twenty-four hours the absorption and lymphatic transport of

154

γ-globulin had virtually ceased in the calf. There was a high correlation between the concentration of lymph γ-globulin and lymph flow, and these authors suggest that the movement of γ-globulin into interstitial fluid during its absorption might cause bulk fluid transfer to the interstitial fluid and thence to the lymphatics.

The absorption of intact protein from the colostrum by the calf is confined to the terminal ileum as shown by histological studies (Smith, 1925; Comline, Roberts & Titchen, 1951*b*). Not only natural substances such as colostrum globulins but also foreign macromolecules, for example radio-iodine labelled polyvinylpyrrolidone (PVP) of mean molecular weight 160000, are absorbed by the intestine of newborn calves and transported in intestinal lymph (Hardy, 1969). Boiled colostrum whey contains factors which enhance intestinal uptake and transport of macromolecules in newborn calves (Balfour & Comline, 1962). Certain simple organic molecules such as lactate, pyruvate and salts of short-chain fatty acids also have this effect. When γ-globulin or PVP is being absorbed from colostrum whey, lymph flow is high and the concentration of the macromolecule in this lymph is low, but a small lymph flow with a high concentration of macromolecule is found when its absorption is enhanced by the presence of these small organic molecules. It is possible that these substances exert their influence on absorption in the lower small intestine by a systemic effect (Hardy, 1969).

Absorption of miscellaneous substances

Glucose

Intestinal lymphatics play no important part in the transport of small water-soluble molecules, such as glucose. For example, Hendrix & Sweet (1917) found no evidence for preferential uptake of glucose into mesenteric lymphatics rather than mesenteric venous blood in dogs, while Hungerford (1959) showed that in fasting rats thoracic duct lymph glucose levels were similar to those of blood plasma, and that during glucose absorption less than 0.3% of the absorbed glucose appears in the lymph. The quantitative difference in transport of glucose by the two routes is explained by differences in flow rates of portal blood and mesenteric lymph. This is strongly supported by the work of Largis & Jacobs (1971) who found that when [^{14}C]glucose was introduced into the jejunum of anaesthetised rats, the specific activities of glucose in the portal venous blood and mesenteric lymph measured throughout the following hour were identical. These activities did exceed that of glucose in peripheral blood (femoral vein). Some glucose, therefore, but only a small amount, is transported from the intestine to the systemic circulation by intestinal lymphatics. Studies of glucose, galactose and fructose in thoracic duct lymph and portal venous blood of dogs after intraduodenal administration of the sugars confirm that lymph is of trivial importance in their transport (Seifert *et al.*, 1975). The concentration of glucose in the lymph of the central lacteal of the villus is probably closely related to that of the fluid initially absorbed from the gut lumen. Lee (1974) using a preparation of mucosal membrane from the dog small intestine found that lymph in the central lacteal is isosmotic with mucosal fluid over a wide range of osmolarity. The glucose concentration in the lacteal was always higher than that of the absorbate of the whole membrane. The explanation for this is not clear though it is possible that some of the initially absorbed fluid loses glucose by metabolism during its transit across the mucosa.

Iron
The idea that the intestinal lymphatics might play a part in iron absorption began with the demonstration by Macallum (1895) that when guinea pigs were fed various preparations of inorganic iron, substantial amounts of iron could be demonstrated in the leucocytes of intestinal villi. Subsequently, Gillman & Ivy (1947) showed uptake of iron in mesenteric lymph nodes in guinea pigs fed ferrous ammonium sulphate, and Gabrio & Salomon (1950) obtained similar results in the horse. These data suggested that lymphatics might be a primary route of iron transport from the gut.

However, Endicott *et al.* (1949), using histochemistry, autoradiography and tracer techniques, concluded that iron in duodenal mucosa and mesenteric lymph nodes behaved as storage iron rather than as iron in transport. Their studies pointed to the portal venous route as the major channel for absorbed iron; in one experiment in a dog, labelled iron, as ferrous ammonium sulphate fed in a single test meal, was found to be transported mainly in portal blood rather than thoracic duct lymph. Peterson & Mann (1952), in a study of rats with intestinal lymphatic fistulae, showed that when labelled iron was given intragastrically, only 2–5% of the total amount absorbed appeared in the lymph. The iron preparations used in these studies were ferric citrate, ferric ammonium citrate, ferrous ascorbate and ferrous gluconate. These findings were confirmed by Everett, Garrett & Simmons (1954), and this group found that the small amounts of iron, which do appear in lymph in rats after a single dose of labelled ferric chloride, are derived secondarily from the bloodstream. Further evidence that iron transport from the gut is chiefly by portal venous blood comes from the work of Reizenstein, Cronkite, Meyer & Usenik (1960). Anaesthetised dogs were fed [^{59}Fe]citrate by stomach tube and the proportion of iron fed which reached the blood via the lymph was estimated to be only about 1.3% of the dose. Thus the bulk of evidence suggests that intestinal lymphatics play no important part in iron absorption.

However, iron is transported within the body bound to protein, and Everett *et al.* (1954) have shown that protein-bound iron given by subcutaneous injection passes almost exclusively into lymphatics. It is unclear why iron, which is presumably complexed with protein while passing through interstitial fluid of the intestinal mucosa during its absorption, should be principally taken up by the bloodstream.

Copper

As with iron, copper absorption does not appear to involve the participation of intestinal lymphatics (Sternlieb, van den Hamer & Alpert, 1967). Fig. 6.19 shows the concentrations of isotopic copper appearing in thoracic duct and portal venous plasma following the administration of labelled copper nitrate to dogs by gastric or duodenal tube. Since portal venous blood flows much faster than intestinal lymph, the amounts of copper

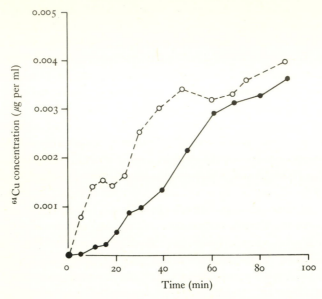

Fig. 6.19. Mean concentration of ^{64}Cu in plasma (O - - - - O) and lymph (●——●) at various times following administration of ^{64}Cu to 3 dogs. 600 μg of ionic copper as copper nitrate was administered in milk to each dog through a peroral rubber tube or by a duodenostomy. (Sternlieb *et al.*, 1967.)

transported by the lymph are trivial compared with those carried by the blood. A single experiment in a patient with a thoracic duct fistula indicated that in man also, the lymphatics play no important part in the transport of copper during its absorption.

Vitamin B$_{12}$

There is a delay of approximately one and a half hours after giving rats labelled vitamin B$_{12}$ before it is found in plasma. This observation prompted Taylor & French (1960) to determine whether this delay was due to lymphatic transport of the vitamin. When a small physiological dose of labelled vitamin B$_{12}$ was given to rats by gastrostomy, only a minor proportion of the absorbed vitamin was found to be transported by thoracic duct lymph. Most of the vitamin B$_{12}$ in the lymph in these circumstances was thought to be derived by filtration from the blood. When a larger dose was given, levels of the vitamin in lymph exceeded those of systemic plasma, suggesting that some might pass directly from the intestine into lymph. Even in this case, it is likely that lymphatic transport of the vitamin is of minor importance when compared with the portal venous route.

Bile acids

Bile acids secreted into the small intestine are extensively reabsorbed, mainly by an active transport mechanism in the lower ileum (see the review by Dowling, 1972). They are transported, bound to albumin, and to some extent to other proteins, in the portal vein to the liver to be resecreted in bile. The lymphatics of the small intestine appear to play little or no part in this conservation of the bile acid pool, since Sjövall & Åkesson (1955) showed that [^{14}C]taurocholate fed to rats could be recovered in bile and none was found in intestinal lymph. Reinke & Wilson (1967), using rats with cannulae in the portal vein, common bile duct, duodenum and intestinal lymph vessel, found that labelled cholic and taurocholic acids are preferentially transported by portal venous blood. Under steady state conditions, the concentration of the bile acids in portal venous blood was twice as great as that in intestinal lymph. This difference is probably due to a greater binding capacity of the blood due to its higher protein content, and this, together with the much faster portal venous flow, accounts for the far greater transport of bile acids by the intestinal blood vessels. Little bile acid ordinarily appears in the systemic blood since liver clearance is very efficient and such small amounts as do appear, may be substantially derived from lymph.

Ammonia

Toxic nitrogenous compounds from the gastro-intestinal tract, particularly the colon, are ordinarily extracted from portal venous blood by the liver, but in liver disease porto-systemic shunts in the splanchnic bed or liver itself deliver these substances to the systemic circulation and brain and this is generally considered a causative factor in the encephalopathy in hepatic disease. The importance of lymph as a potential bypass of the liver is obvious and the interesting observations by Lee

& Duncan (1968) indicating that elevation of portal venous pressure directs more absorbed fluid, and presumably solute, into lymph are relevant to this problem.

There is, however, no reason to expect that ammonia produced in the normal gastro-intestinal tract should be preferentially transported by lymph. Denis (1968) showed that ammonia is present in lymph collected from the cervical trunk, the thoracic duct and the cisterna chyli of dogs, and levels of ammonia in these 'central' channels were found to be comparable with those of arterial and venous blood, both in the normal state and in hyperammonaemia. The tissues which produce the greatest amounts of ammonia are the kidney and the large intestine, while the liver is an ammonia acceptor organ. Lymph and blood draining the kidney and large intestine may have higher concentrations of ammonia than does 'central' lymph and blood. One would expect that concentrations of ammonia in lymph and blood would rapidly equalise by diffusion.

Particulate material

It is possible that small amounts of particulate material may enter the body by intercellular passage through the intestinal mucosa (Volkheimer, Schultz, Lindenau & Beitz, 1969). This has been termed 'persorption' and the phenomenon has been studied with starch granules and particles of metallic iron. Volkheimer and his associates believe that a 'kneading' action allows particles of up to 100 μm diameter to enter the lamina propria of the intestinal mucosa near the apex of the villus. Metallic iron particles of up to 52 μm were identified in both portal venous blood and thoracic duct lymph of dogs after the animals were fed 200 g of iron powder suspended in milk and cream. This phenomenon is rather curious and its quantitative significance is very dubious. Large foreign particles should be expected to enter intestinal lymph vessels in preference to mucosal blood vessels, and it is remarkable that such large particles should appear at all in portal venous blood.

Drugs and foreign substances

There is comparatively little information on this subject. If a drug is efficiently metabolised in the liver, the lymphatic route could be an important bypass delivering a substantial amount of it to the systemic circulation, especially if the substance partitions in favour of the lymphatic route at the site of absorption. Lipophilic drugs and foreign substances might become incorporated into chylomicra and thus be delivered at high concentration at sites where chylomicra are metabolised. Since many drugs and other foreign compounds are bound to protein in plasma, the passage of such drugs through the interstitial space of the intestinal mucosa could allow drug–protein binding with subsequent lymph transport to a greater or lesser degree. In this section both hydrophilic and lipophilic foreign compounds are considered.

Studies of mesenteric lymph and blood plasma levels of *p*-amino salicylic acid (PAS) in rats after duodenal administration of the drug suggest that direct transport of PAS by lymphatics occurs (De Marco & Levine, 1969). These authors have also demonstrated by fluorescence microscopy that tetracycline appears in the central lacteal of villi, in the area of the small intestine which is absorbing the drug. Despite the fact that these drugs are, to some extent, transported in lymph, the quantitative importance of the intestinal lymphatic route in delivering the drug to the circulation is small due to the relatively slow rate of intestinal lymph flow as compared with portal blood flow. Under certain physiological and pathological circumstances, however, the lymphatic route may assume greater importance. For example, when glyceryl tripalmitate was given by mouth, this substance acting as a lymphagogue, increased the quantities of the drugs transported in the lymph by increasing lymph flow, while reduction in portal blood flow brought about by ligation of the superior mesenteric vein increased lymphatic transport of PAS.

Concurrent feeding of a lipid may enhance the lymphatic transport of a drug by mechanisms other than a pure lymphagogic effect. For example, Gianinna, Steinetz & Meli (1966), studying the pathway of absorption of orally administered ethynyl oestradiol-3 cyclopentyl ether in the rat, found that the drug given as an aqueous suspension was mainly carried in the portal vein, but when the drug was given in sesame oil, the lymphatic route assumed some importance in its transport. (The predominant fatty acids of sesame oil glycerides are oleic and linoleic.) Addition of glyceryl mono-oleate to the sesame oil augmented this effect. These authors postulated a physical or chemical interaction between the drug and its lipid vehicle to account for the effect. The drug is a highly lipophilic steroid derivative and it seems possible that it might be partly transported in chylomicra formed during digestion and absorption of the oil. The inclusion of the long-chain monoglyceride in the meal would facilitate micellar solubilisation of the drug in the lumen of the small intestine in a way similar to its action in enhancing the solubility of cholesterol in bile acid micelles. It is not clear, however, why this effect should specifically enhance lymphatic transport of the drug.

Steroid hormones less hydrophobic than ethynyl oestradiol-3 cyclopentyl ether appear to be transported mainly by the portal venous route. In rats, only trivial amounts of 17-methyl[^{14}C]oestradiol (Bocklage *et al.*, 1953) and of 17 α-methyl[^{14}C]testosterone (Hyde, Doisy, Elliott & Doisy, 1954) are transported by the lymphatic pathway. Similarly the ^{14}C of cortisone-4[^{14}C]acetate was not absorbed by intestinal lymphatics of rats after the hormone was given intragastrically. (Bocklage, Doisy, Elliot & Doisy, 1955), and Hellman, Bradlow, Frazell & Gallagher, (1956) found that in man, orally administered labelled hydrocortisone and testosterone were not transported to a significant degree in thoracic duct lymph.

It might be expected that digitoxin, a lipid-soluble drug with low water solubility and a steroid ring, might be transported to a substantial extent by intestinal lymph. However, Oliver, Cooksey, Witte & Witte (1971) gave [^{3}H]digitoxin either intragastrically or intraduodenally to anaesthetised dogs and found the drug to be rapidly and uniformly absorbed from the duodenum, but travelling by the portal vein and not by intestinal lymph. This finding is confirmed by the work of Beermann & Hellström (1971) who found that, over forty-eight hours, lymph drainage through a thoracic duct fistula accounted for only about 4% of a dose of labelled digitoxin or digoxin given intragastrically to rats. Only trivial amounts of proscillaridin A, a cardiac glycoside derived from squill, were found in human thoracic duct lymph after the drug was fed to two human subjects (Andersson *et al.*, 1977). Thus it appears that lymph does not constitute an important pathway for the transport of cardiac glycosides from the intestine. Neither is lymph an important route for transport of bromsulphthalein from the intestine in rats (Vanlerenberghe, Trupin, Nguyen & Rose (1972).

Sieber, Cohn & Wynn (1974) have studied the lymphatic transport of a large series of labelled foreign and natural compounds given to rats with thoracic duct fistulae. Radioactivity appeared in the lymph in small quantities after all the compounds were given but the portal route appeared to be more important in intestinal absorption of benzene, benzoic acid, aniline, *p*-aminobenzoic acid, salicylic acid, phenanthrene, oestradiol, testosterone, digoxin, hexanoic acid, hexylamine, hexanol, antipyrine, isoniazid and caffeine. Octadecanoic (stearic) acid, octadecanol, cholesterol and p,p^1-DDT, however appeared to be selectively absorbed into intestinal lymph. The DDT, which has a relatively high lipid solubility, appeared to be located in the lipid core of chylomicra. Pocock & Vost (1974) have found in rats that 60% of a 100 nmole dose of ^{14}C-labelled DDT can be recovered in 12 hours in thoracic duct lymph as a component of chylomicra. Using ^{3}H-labelled triglyceride and ^{32}P-labelled phospholipid as markers of chylomicron core and coat respectively, they showed that 97% of the DDT is present in the triglyceride core. This DDT in the chylomicron appears to be able to transfer to higher density serum proteins in the circulation. It is interesting that absorption of DDT into lymph is enhanced by concomitant fat feeding. A similar phenomenon may account for the observation that tetrachlorethylene, a drug given to treat hookworm infections, which is not ordinarily absorbed, causes toxic effects when fed with a fat meal. There are other examples of enhanced absorption of non-polar compounds during fat absorption (see Chapter 7) and this interesting phenomenon deserves further attention.

Summary

The major route of transport of water and water-soluble substances from the small intestine is the portal vein and its tributaries. The lymphatic system, however, does appear to contribute to the transport of fluid absorbed from the intestine. The partition of this fluid between venous and lymph vascular channels is variable and depends on several factors. These include the tonicity of the fluid in the gut lumen, intestinal motility and distension pressure and pressure in the portal venous system; it should be emphasised that as yet it is quite unclear what influence these various factors have in the physiological situation. It seems reasonable to regard the lymphatics as an overflow system with respect to excess mucosal fluid and as an alternative pathway for fluid transport when portal venous pressure is raised.

Certain macromolecules, proteins and peptides, which are absorbed from the intestine or are synthesised locally in the mucosa are transported by lymphatics. For example, in newborn animals, some intact absorbed protein is conveyed to the general circulation by lymph, and lipoproteins, newly synthesised by small intestinal epithelium, are also delivered by this route. Intestinal lymph is a major route of delivery of IgA, synthesised in the gut mucosa, to the bloodstream. Drugs and certain other foreign substances are conveyed to the general circulation in part, at least, by lymph. Binding of such foreign substances to proteins in the interstitial fluid might play a part in directing them towards the lymphatics, while the concurrent administration of a lymphagogue, such as long-chain triglyceride, with the test substance enhances the importance of the lymphatic route by augmenting lymph flow. The transport of drugs and foreign substances by lymphatics is interesting and requires further study.

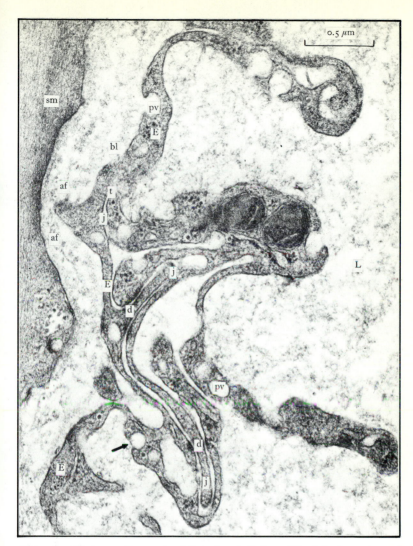

Plate 2.2. Electron micrograph of lacteal endothelium (E) of fasting guinea pig illustrating marked complexity of junctional complex (j) with considerable interlocking and overlapping of endothelial processes. There are several desmosome-like (d) points of attachment and a probable tight junction (t). Pinocytotic vesicles (pv) are prominent and one appears to have a diaphragm (arrow) stretching across its stoma. The basal lamina (bl) is quite inapparent and can hardly be distinguished from precipitated proteins in the lumen of the lacteal and in the extracellular spaces. A portion of smooth muscle (sm) can be seen adjacent to the lacteal endothelium. Anchoring filaments (af) may attach lacteal endothelia to adjacent smooth muscle. (From Dobbins, 1971.)

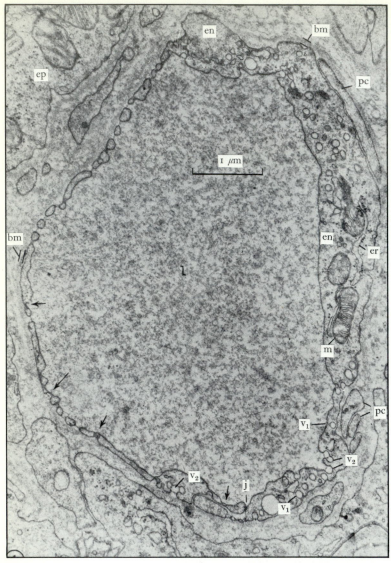

Plate 3.1. Transverse section of a fenestrated capillary in the intestinal mucosa. Abbreviations: ep, intestinal epithelium; bm, basement membrane; en, endothelium; pc, pericyte; er, endoplasmic reticulum; m, mitochondrion; v_1, flask-shaped vesicle; v_2, apertured vesicle; j, junctional element. Arrows show areas which suggest plasmalemma fusion forming fenestrae. (From Clementi & Palade, 1969.)

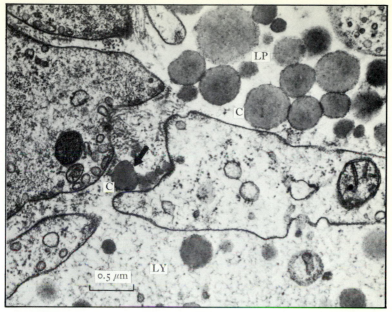

Plate 3.2. Chylomicra (C) entering a lymphatic (LY) through an inter-endothelial gap (arrow); LP, lamina propria of intestinal villus. (Courtesy of MTP Press Ltd.; Sabesin 1976.)

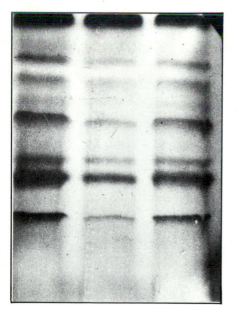

Plate 3.3. Starch gel electrophoresis of rat plasma and intestinal lymph. Two specimens of plasma flank a specimen of cell-free lymph. Gels were run horizontally for 3 hours at 8 V/cm between cooling plates. The starch was made up in a buffer composed of 0.076 M tris (tris hydroxy methylamino methane) and 0.005 M citric acid (final pH 8.8). The buffer in the electrode vessels was 0.3 M boric acid and 0.05 M NaOH (pH 8.1). The gel was stained with naphthalene black. Anode at top.

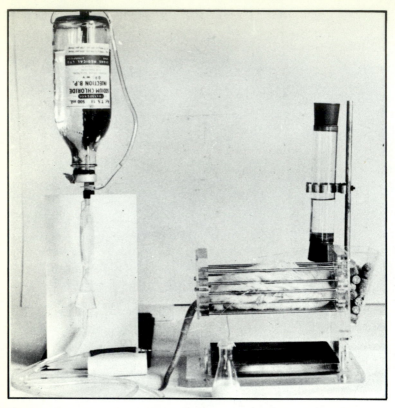

Plate 4.1. Rat with intestinal lymph fistula and femoral intravenous infusion in a Bollman cage.

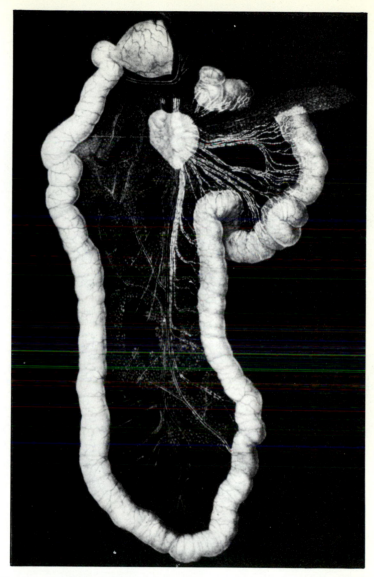

Plate 7.1. Transport of absorbed fat by lacteals from the duodenum of the rabbit. At the upper left, the bile duct enters the duodenum just below the pylorus. White lymphatics are readily seen only distal to the entry of the pancreatic duct at the lower right in the picture. (From Bernard, 1856.)

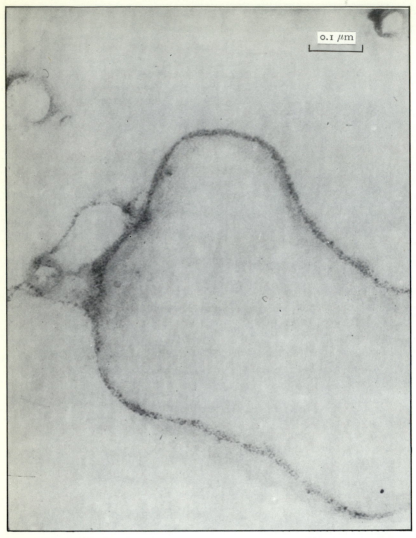

Plate 7.2. Chylomicron coats prepared by 'oiling out' triglyceride from intact chylo-micra. Chylomicron 'ghosts' prepared by sedimentation at 100000 g for 20 min. (Zilversmit, 1965.)

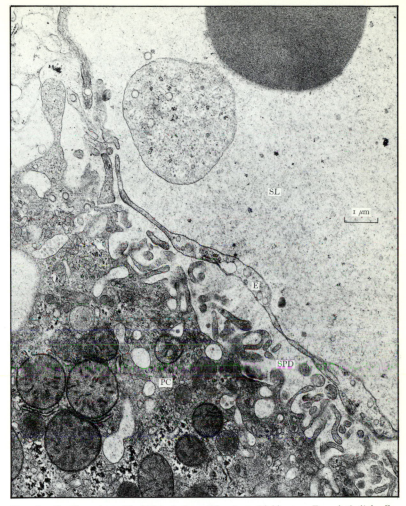

Plate 8.1. Rat liver sinusoid. Abbreviations: SL, sinusoidal lumen; E, endothelial cell; SPD, perisinusoidal space of Disse; PC, parenchymal cell.

Plate 9.1 Intestinal lymphangiectasia in a small intestinal biopsy. The central cavity has an endothelial cell lining.

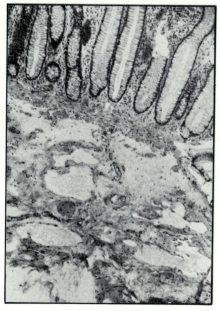

Plate 9.2 Lymphangiectasia of the large intestine. Mucosal biopsy from the recto-sigmoid area showing dilated lymphatics and oedema in the submucosa. (Schaefer *et al.*, 1968.)

7 TRANSPORT OF ABSORBED LIPIDS

Of all the functions of the lymphatic system of the gastro-intestinal tract, the best known is its participation in the process of fat absorption. It was a post-prandial rise in the concentration of particulate fat in the mesenteric lymph of a dog which enabled Asellius to make his historic demonstration of intestinal lymph vessels and their contents. Two centuries later, Claude Bernard used this observation to demonstrate that pancreatic secretion is required for the intestinal absorption of fat (Plate 7.1).

In view of the great importance of intestinal lymphatics in the transport of the heterogeneous group of substances classified as lipids, an introductory section of this chapter briefly outlines present concepts of fat digestion and metabolism in the intestine. The remainder of the chapter reviews the role of alimentary tract lymphatics in fat absorption.

Dietary lipids

Lipids form an important part of the diet in most species. Although human eating habits vary greatly from country to country, most Western diets contain amounts of fat which contribute between 20 and 40% of the total calorie intake. Dietary lipids comprise a variety of substances found in several different foods of animal and vegetable origin. They are insoluble in water, but readily dissolve in non-polar solvents such as ether, chloroform or benzene, and are generally esters of fatty acids. Sterols which occur in the diet, such as cholesterol, may be present as the free form or esterified with fatty acids. In moderate amounts lipids enhance the palatability of a diet, and offer a high calorie yield.

This section considers the lymphatic transport of the various species of lipid in the diet, i.e. triglycerides, sterols, phospholipids, and the fat-soluble vitamins. Lymphatic transport of drugs and foreign compounds of varying degrees of water and lipid solubility was considered in Chapter 6. An important feature of the absorption of dietary lipids which is emphasised in each section is that exogenous (dietary) lipids are to some extent diluted by endogenous lipid derived chiefly from bile and shed mucosal epithelium. Clearly, this dilution effect is of greatest importance where dietary intake of a lipid type is relatively small, as is the case for cholesterol.

Triglycerides constitute the largest proportion of dietary lipids and body fat stores; by virtue of their high calorific value and the absence of associated water, they form a compact store of energy for the organism. Triglycerides are fatty acid esters of glycerol. Dietary triglycerides are derived from animal body and milk fats, and from vegetable oils; they contain a wide variety of fatty acids. In some glycerides, such as glyceryl trioleate (triolein), the same fatty acid species esterifies all three alcohol groups of glycerol. Many triglycerides, however, have mixed fatty acid composition. The fatty acids of naturally occurring triglycerides may be divided into those which are saturated and those which have one or more double bonds in their chain, the mono- and poly-unsaturated fatty acids. Among these unsaturated fatty acids, there exist a few species, such as linoleic acid, which appear to be essential for normal nutrition of experimental animals as well as man. In general, the fatty acids of naturally occurring glycerides have long chains of even numbers of carbon atoms of which the C_{16} saturated acid, palmitic, is found in large amounts in many fats. Its homologue, stearic acid (C_{18}), is found somewhat less frequently. Of the unsaturated group, oleic acid (C_{18}), which has one double bond, is the most prominent. Milk and its products contain glycerides with a small proportion of short chain fatty acids, such as butyric acid. In an average diet in Western countries, an individual will take approximately 70–100 g of triglyceride per day.

An overall view of lipid digestion and absorption

Intragastric digestion

It is unlikely that, in adult life, extensive digestion of lipids takes place in the stomach. Lipolytic activity towards tributyrin was identified by Schønheyder & Volqvartz in human gastric juice in 1946. This enzyme had a pH optimum of 5.5 for tributyrin hydrolysis; the optimum pH for hydrolysis of longer-chain triglycerides was higher. The overall hydrolytic activity of this enzyme towards long-chain triglycerides which constitute the bulk of normal dietary glycerides was very small. Its importance might be greatest in the newborn since appreciable amounts of short-chain triglyceride are present in human and cows' milk. Milk itself, however, contains lipases which are capable of hydrolytic activity under conditions prevailing in the stomach (Hernell & Olivecrona, 1974). There has been recent interest in another source of intragastric lipolytic activity, namely the serous glands of the tongue of the rat, secreting the so-called 'lingual lipase' (Hamosh & Scow, 1973). In man, these same investigators have identified a similar pharyngeal lipase having a pH optimum of 5.4 and these enzymes may be responsible for preliminary lipolysis in the stomach (Hamosh, Klaeveman, Wolf & Scow, 1975). The split products of lipolysis together with proteins of dietary and gut origin (Meyer, Stevenson & Watts, 1976) would facilitate preliminary emulsification of dietary lipids in the stomach. Long-chain triglycerides may be hydrolysed to a slight extent by pancreatic lipase within the stomach as a result of reflux of duodenal contents. This reflux is probably small in normal subjects, and pancreatic lipase which has a pH optimum close to neutrality in the presence of bile salts would be able to hydrolyse very little glyceride in the acid conditions of the stomach. Furthermore, the action of pancreatic lipase on its insoluble substrate is greatly enhanced by fine dispersion of the glyceride which is largely accomplished by the bile salts within the intestinal lumen, whereas fats in the stomach are only poorly dispersed. Since optimal conditions for lipolysis are not found in the stomach, glyceride hydrolysis is mainly accomplished after chyme has entered the small intestine.

167

Transport of absorbed lipids

Digestion and uptake in the small intestine

Fat entering the small intestine stimulates, by predominantly humoral mechanisms, the release of bile and pancreatic juice. These two secretions play a fundamental role in the preparation of lipids for absorption. The fat in gastric chyme is in the form of a very coarse emulsion which is produced mainly by the mechanical action of the stomach and by certain emulsifying agents such as proteins, present in the diet and in gastro-intestinal secretions (Meyer *et al.*, 1976). From the point of view of fat digestion, bile contains two important constituents; bile salts and phospholipids, both of which promote the emulsification of lipids within the small intestinal lumen. The conversion of the phospholipid lecithin to lysolecithin within the gut by the action of phospholipase enhances this property of the phospholipids. Other components of the diet, such as proteins, assist in the stabilisation of the emulsion of fats in the aqueous phase of small intestinal content.

The action of pancreatic lipase in hydrolysing the 1- and 3-ester bonds of triglyceride is rapid, yielding monoglycerides and fatty acids. In the presence of bile salts in the small intestine, pancreatic lipase associates with a small protein which is secreted in pancreatic juice. This 'colipase' enables lipase to digest triglyceride in the presence of conjugated bile salts at the prevailing pH of the upper small intestine – approximately 6.5 (Borgström, 1975). The hydrolysis products of triglyceride dissolve in micelles of bile salts to form mixed micelles in the lumen of the small intestine (Hofmann & Borgstrom, 1962). The micellar solubilisation of monoglycerides and fatty acids removes them from the oil–water interface at which pancreatic lipase acts, and this enhances digestion of the remaining triglyceride. Some monoglyceride and fatty acid, however, remains associated with the oil phase, i.e. the undigested triglyceride. The absorption of the products of lipolysis probably occurs from molecular solution in the aqueous phase. The micellar solution acts as a convenient 'store' of these compounds, since the aqueous solubility of monoglycerides and long-chain fatty acids is very small. As mixed bile salt–lipid micelles approach the luminal surface of the enterocyte considerable resistance to diffusion of these aggregates is offered by an 'unstirred layer' of water adjacent to the cell membrane (Wilson, Sallee & Dietschy, 1971). It is likely that little or no unhydrolysed long-chain triglyceride is absorbed by the enterocyte (Cardell, Badenhausen & Porter, 1967).

An essential function which mixed micelles appear to serve is the solubilisation of highly non-polar solutes such as the sterols and fat-soluble vitamins. These dissolve in the hydrophobic interior of micelles and the presence of bile acids in the small intestine is required for absorption of such solutes.

Some passive absorption of bile acids occurs in the upper small intestine. All the contents of the mixed micelles, however – the monoglycerides, fatty acids and non-polar solutes – pass into the absorptive epithelium by a non-energy dependent process (Strauss, 1966). This statement requires the qualification that while no energy-dependent carrier mechanism for lipids is recognised in the epithelial cell, energy-dependent metabolic transformation of lipids within the cell will influence the total transfer of these lipid species across the cell. Furthermore, metabolic activity may influence permeability or selectiveness of the cell membrane to different lipids. An intracellular fatty acid-binding protein may be important in facilitating the transport of fatty acid across the cell membrane and through the cytosol (Ockner, Manning, Poppenhausen & Ho, 1972). On entry into the mucosal cell, monoglycerides and fatty acids are re-esterified to triglycerides. Two mechanisms of re-esterification are recognised. Fatty acids, activated to coenzyme A derivatives (thiolesters) interact with α-glycerophosphate to form lysophosphatidic acid which yields phosphatidic acid after a second acylation. Phosphatidic acid is hydrolysed by a phosphatase to yield a diglyceride which is acylated again with a coenzyme A derivative of a fatty acid to yield a triglyceride. A more important pathway involves direct acylation of absorbed monoglyceride to triglyceride by interaction with thiolesters of fatty acids. However some hydrolysis of absorbed monoglyceride to glycerol and free fatty acid may occur in the mucosal cell by the action of a monoglyceride lipase. The glycerol so released is probably partly transported in portal venous blood and partly metabolised within the

Transport of absorbed lipids

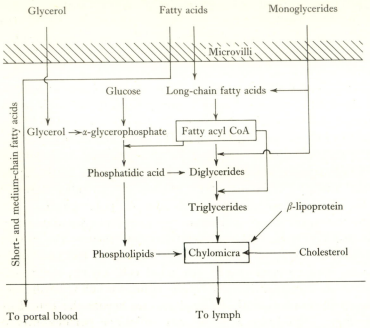

Fig. 7.1. Uptake and resynthesis of the products of triglyceride hydrolysis by the intestinal epithelial cell.

mucosal epithelial cell. The synthetic enzyme systems responsible for the formation of triglyceride in the mucosal cell are located in the smooth endoplasmic reticulum, and the Golgi region appears to play an important part in the accumulation of the resynthesised triglycerides. Subsequently the triglycerides associate with phospholipids and proteins as chylomicra which leave the mucosal cell to be transported in the intestinal lymph. These processes of triglyceride uptake by the mucosal epithelium of the small intestine are summarised in Fig. 7.1.

Absorption of medium- and short-chain triglycerides

The description of triglyceride hydrolysis and resynthesis given above applies to the bulk of dietary triglyceride, i.e. the long-chain type. There are important differences between the handling of medium- and short-chain triglycerides (ten carbon

170

atoms chain-length and less) and the long-chain types. First, pancreatic lipase is more active in the hydrolysis of medium- than long-chain triglycerides and the intraluminal hydrolysis of medium-chain triglycerides proceeds very rapidly in the small intestine. Secondly, medium-chain fatty acids are less readily re-esterified to triglyceride in the mucosal cell, possibly because they have less affinity for the esterifying enzymes. They are generally absorbed into portal venous blood bound to albumin and therefore reach the liver more directly than long-chain fatty acids (see below). A third difference between the long- and medium-chain triglycerides is that the latter are to some extent absorbed intact by the mucosal epithelium and may be subsequently digested by a mucosal lipase in the absorptive cell (Playoust & Isselbacher, 1964). These phenomena can be utilised in the dietary management of patients suffering from conditions where there is defective intraluminal lipolysis in the small intestine or where lymphatic transport of absorbed lipid is impeded.

Mucosal protein synthesis and chylomicron formation
The mechanism by which chylomicra gain a surface protein coat within the mucosal cell is not yet fully understood. It is possible that the oil droplets of resynthesised triglyceride are formed in a proteinaceous medium within the membrane-bound cisternae of the cell and a layer of lipophilic protein forms around each droplet (Cardell *et al.*, 1967). Hatch, Aso, Hagopian & Rubenstein (1966) have found that the amino acids of chylomicron protein include a considerable proportion of non-polar types, and this confers a degree of lipid solubility on the proteins and favours their association with lipids. The proteins associated with chylomicra are probably formed at the ribosomes of the rough endoplasmic reticulum of the epithelial cell, and pass through communications between the cisternae of the rough and smooth endoplasmic reticulum into the Golgi saccules where they associate with the resynthesised triglycerides.

Several attempts have been made to throw light on the part played by lipoprotein synthesis in fat absorption and transport in mucosal epithelium. The rare disease of a-β-lipoproteinaemia

171

is a clinical condition whose features point to an obligatory role of β-lipoproteins, synthesised in the intestinal mucosa, in the formation of chylomicra. This is an autosomal recessive condition in which there are several abnormalities of lipid transport and metabolism, a major abnormality being the absence of low density lipoproteins from the plasma (Scanu *et al.*, 1974). It presents in childhood as a steatorrhoea. After fat ingestion, post-prandial lipaemia fails to occur and the upper small intestinal mucosal epithelium can be shown to be engorged with neutral lipid. Other features of a-β-lipoproteinaemia include retinitis pigmentosa and a progressive cerebellar and posterolateral column degeneration which leads to weakness and ataxia. One other characteristic feature of the condition is the presence of acanthocytes, an unusual form of crenated red cells whose shape may reflect an abnormality of the lipoprotein membrane of the cell. A specific apoprotein (apolipoprotein-B) present in β-lipoproteins and chylomicra is absent from the serum of patients with this disease (Gotto, Levy, John & Fredrickson, 1971). For a review of this subject see Kostner (1976).

The mucosal defect in handling fat appears to occur relatively late in the process of fat transport across the mucosa, since [^{14}C]palmitate given by mouth is readily incorporated into mucosal triglyceride. It is concluded, therefore, that failure of β-lipoprotein synthesis, or of its coupling with lipids is responsible for the observed effects.

What is of great interest in this condition is that although the steatorrhoea of the child with the condition lessens with age, partly due to dietary avoidance of fat or due to the use of medium-chain triglyceride, there is nonetheless evidence that a significant quantity of dietary fat can be absorbed as the patient matures and it seems possible that portal venous transport of long-chain fatty acids (i.e. greater than fourteen carbon atoms chain-length) may occur in these patients; this is a route normally reserved principally for short chain fatty acids. As will be seen later in respect to other lipids, the portal venous route may have a minor but facultative role in transport of all lipid species.

The observations on a-β-lipoproteinaemia have led several workers to attempt to create an experimental model of the condition. Isselbacher & Budz (1963) showed that puromycin, a protein synthesis inhibitor which is thought to achieve its effects by displacing peptide chains from the ribosomes prior to their final assembly as complete proteins, greatly inhibits the incorporation of [^{14}C]leucine into rat intestinal mucosal proteins and, in particular, into the proteins of chylomicra synthesised in these cells. It seemed likely, therefore, that puromycin might create an experimental inhibition of fat transport. Studies by Sabesin & Isselbacher (1965) have suggested that a situation akin to human a-β-lipoproteinaemia does occur in rats given puromycin. In their study, such animals fed corn oil showed accumulation of lipid droplets within mucosal epithelial cells. Little fat was seen in villus lymphatics, and where normal rats fed 1.5 ml of corn oil will show a mean plasma triglyceride concentration of about 380 mg per 100 ml six hours after feeding, puromycin-treated animals scarcely showed any rise in serum triglyceride above the levels observed in fasting animals. The lipid in mucosal epithelial cells of puromycin-treated animals was found to be triglyceride, suggesting that the production of enzymes of triglyceride resynthesis is relatively insensitive to puromycin, or that sufficient enzyme is present in the cell to render the treatment ineffective in halting triglyceride synthesis. Other evidence that puromycin does not affect triglyceride resynthesis came from studies of [^{14}C]palmitate incorporation into mucosal triglyceride.

This experimental model promised to be of great value in studying various aspects of mucosal handling of absorbed lipid, and has been so used by various workers. For example, Kayden & Medick (1969) have used it to study the partition of fatty acid transport between mesenteric lymphatics and the portal vein. However, it has been questioned whether the model creates a situation at all comparable with a-β-lipoproteinaemia. The effects of various inhibitors of protein synthesis on lipid absorption have recently been examined. Redgrave (1969), studying the uptake of lipid by the thoracic duct of rats given a bile salt-micellar solution of monoglyceride and fatty acid by duodenal infusion, found that actinomycin in doses which inhibited intestinal mucosal protein synthesis, as judged by

[14]C-labelled amino acid incorporation, did not affect lipid transport into lymph. Puromycin, the compound most often used to produce the model, was re-examined by Redgrave & Zilversmit (1969). Orally administered fat was absorbed only slowly from the gastro-intestinal tract of rats given puromycin, but this was found to be largely due to impairment of gastric emptying. When lipid was given into the duodenum it was efficiently absorbed in puromycin-treated rats. Lymph flow, however, was reduced by puromycin treatment and, despite an increase in lymph triglyceride concentration, the overall triglyceride transport in the thoracic duct of puromycin-treated animals was lower than in control animals. This reduction seemed to be mainly due to the reduction in lymph flow produced by puromycin. Such reduction has been observed with other inhibitors of protein synthesis.

The results of Redgrave & Zilversmit suggest that, in the intact animal, inhibitors of protein synthesis may exert their action on fat absorption by several mechanisms, such as a delay in gastric emptying, a reduction in intestinal lymph flow of unknown cause and possibly a direct inhibition of β-lipoprotein synthesis. Yet another possible mechanism for interference in fat absorption by puromycin is through an inhibition of membrane protein synthesis as described by Friedman & Cardell (1972), who have found electron microscopic evidence in rat intestinal epithelium of dramatic reduction of intracellular membrane material within one hour of treatment. Such an effect might well interrupt fat transport through the cell. Caution, therefore, is required in making a close comparison between the effects produced by puromycin treatment in experimental animals and human a-β-lipoproteinaemia.

Glickman, Kirsch & Isselbacher (1972), attempting to clarify this matter studied the size of lymph chylomicra during inhibition of protein synthesis with acetoxycycloheximide. This study showed that inhibition of protein synthesis *did* reduce fat absorption and produced a marked and sustained increase in chylomicron size for up to four hours after lipid infusion into the duodenum of rats. This latter observation is in keeping with the hypothesis that fat absorption is markedly influenced by concurrent mucosal protein synthesis if one regards the

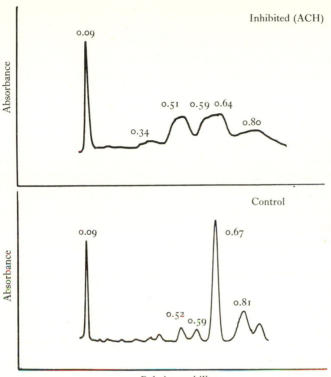

Fig. 7.2. Effect of acetoxycycloheximide (ACH) on chylomicron apoproteins. Chylomicra were prepared from control and ACH-treated rats and subjected to sodium dodecyl sulphate acrylamide electrophoresis. The densitometric scans of gels stained with Coomassie Blue are shown. The relative mobilities (Rf) are indicated above each peak. (Glickman & Kirsch, 1973.)

enlarged chylomicra as an attempt to package lipid more efficiently with limited supplies of apoprotein. Control experiments showed that in normal animals a transient increase in chylomicron size occurs during maximal triglyceride absorption, a phenomenon previously described by Fraser, Cliff & Courtice (1968). Subsequently, Glickman & Kirsch (1973), using electrophoresis in sodium dodecyl sulphate polyacrylamide gels to separate the apoproteins of chylomicra, showed that in untreated rats there was no significant difference in these apoproteins derived from chylomicra of different sizes, but in animals treated with acetoxycycloheximide the proportion of a major apoprotein which appeared to be immunologically related to high density lipoprotein was greatly reduced (Fig. 7.2). Thus the balance of evidence supports the view that lipoprotein synthesis in the enterocyte has a marked influence on the absorption and transport of long-chain triglyceride.

Lymph flow and fat absorption

Fat feeding has a well recognised lymphagogic effect. Tasker (1951) observed a marked rise in intestinal lymph flow in rats following intragastric administration of olive oil. A similar effect has been observed by several other workers including Borgström & Laurell (1953) who noted that there was no fall in the protein concentration of the lymph during the period of enhanced flow, and therefore fat absorption is accompanied by a large increase in the lymphatic transport of protein (Fig. 7.3). When saline was given by mouth to rats lymph flow also rose, but this lymph was diluted with respect to protein. An interesting aspect of this augmentation of lymph flow during fat absorption is the time course of the changes, since these authors found that peak lymph flows occurred about two hours before peak concentrations of lymph lipid were attained, an observation later confirmed by Simmonds (1955). Wollin & Jaques (1973) have shown that olive oil given to rats (2 ml per kg body weight) causes a marked escape of plasma proteins labelled with Evans Blue into intestinal lymph while equivalent

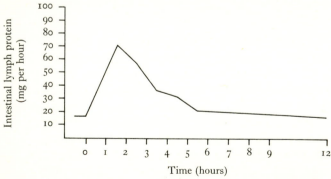

Fig. 7.3. Intestinal lymph protein flow before and after peroral administration of 1 ml of corn oil to rats. The line marks the mean values from six rats (Borgström & Laurell, 1953.)

volumes of 50% glucose did not have this effect. Thus the probable explanation for the high protein flow in intestinal lymphatics during fat absorption is an enhanced blood flow to the intestinal mucosa or enhanced intestinal capillary permeability with increased filtration of plasma protein into the interstitial space of the villi. This would also explain the rise in lymph flow, and the dissociation between lymph flow and lymph lipid concentration could be explained by an alteration in mucosal blood flow brought about by a neural or humoral mechanism initiated during the early phases of fat absorption. Fara, Rubinstein & Sonnenschein (1972) have found that corn oil, L-phenylalanine or acid instillation into the duodenum of the cat produces a selective mesenteric vasodilatation which can also be produced by intravenous infusions of low doses of cholecystokinin and secretin. The mechanism by which the nutrients given in these experiments produce the haemodynamic effects is not certain though they are powerful stimulants to endogenous cholecystokinin release. It is conceivable that cholecystokinin, possibly in association with other gastrointestinal humoral agents, is involved in the enhanced lymph flow and plasma protein leak which follows fat feeding. The participation of 5-hydroxytryptamine in the intestinal vascular responses to gastro-intestinal hormones has been proposed by Biber, Fara & Lundgren (1973b).

An alternative possibility is that during fat absorption intestinal motility increases, which is also a recognised effect of cholecystokinin (Fara *et al.*, 1972), and this might enhance capillary filtration and lymph flow. This is less likely in view of studies by Simmonds and his colleagues. In 1957, Simmonds showed that the rise in lymph flow and fat output of a thoracic duct fistula in rats receiving an infusion of lipid into the small intestine was unaffected by enhancing or depressing the motility of the intestine with cholinergic or anti-cholinergic agents. Bennett, Shepherd & Simmonds (1962) infused emulsified coconut oil into the duodenum of unanaesthetised rats which had thoracic duct fistulae and observed that morphine injected as a single dose into the duodenum produced intestinal spasm which caused regurgitation of about half the lipid into the stomach and thereby reduced lymph transport of lipid; the absorption of the fat retained in the small intestine was not affected. Intestinal motility, therefore, plays little part in the augmentation of lymph flow and protein transport in the thoracic duct during fat absorption, and an alteration in mucosal blood flow, possibly with an accompanying change in intestinal capillary permeability, remains the likely explanation for the lymphagogic effect of fat feeding.

Wollin & Jaques (1973) have demonstrated that during fat absorption plasma protein passes in excess of basal amounts into intestinal lymph and this has been confirmed by Barrowman & Turner (1976) who have shown that the transport of the absorbed lipid by the lymphatic route is not necessary for this effect since it occurs when medium-chain triglycerides, which are mainly transported in portal venous blood, are being digested and absorbed (see Fig. 7.4). Free fatty acids introduced into the small intestine also seem to increase lymph flow, though their effect seems to be less pronounced than that of triglycerides. It is probable that lipids, or the products of lipid digestion, initiate a local vascular change in the gut mucosa resulting in enhanced filtration of plasma protein into the interstitial space of the gut and thence to intestinal lymph. The mechanism by which fats enhance intestinal blood flow is as yet unclear though the studies of Fara *et al.* (1972) and Fara & Madden (1975) suggest that humoral mechanisms, possibly involving cholecystokinin and secretin are at least partly involved (see Chapter 3).

Wollin & Jaques (1974) have found that there is a substantial increase in the activity of diamine oxidase in intestinal lymph of rats after olive oil feeding and have suggested that this may reflect a release of histamine which mediates the vascular changes presumed to be responsible for the increased lymph flow: some further support for this notion comes from the demonstration that treating rats with a combination of an H_1-receptor antagonist, pyrilamine, and an H_2-receptor antagonist, burimamide, blocks plasma protein escape from the blood to intestinal lymph during fat feeding (Wollin & Jaques, 1976). This interesting possibility needs further investigation.

While fat absorption is associated with increased lymph flow and protein transport in lymph, it may be that the converse

178

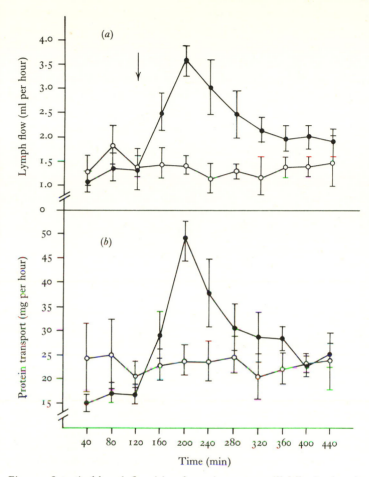

Fig. 7.4. Intestinal lymph flow (*a*) and protein transport (*b*) following intraduodenal administration (marked by the arrow) of 0.5 ml medium-chain triglyceride (closed circles) or 0.5 ml 0.9% NaCl (open circles). Results are means and standard error of mean for groups of five rats. (Turner & Barrowman, 1977.)

is also true, i.e. that augmented lymph flow increases fat absorption and transport. A study by Baraona & Lieber (1975) has shown that the administration of ethanol to rats not previously given ethanol enhances the flow of intestinal lymph. Augmented amounts of proteins and lipids are present in this lymph and some of this excess lipid was shown to be dietary in origin (Fig. 7.5). There is no clear explanation for these

179

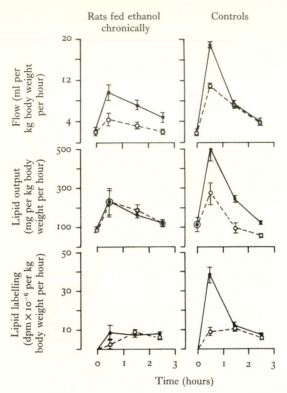

Fig. 7.5. Time course of intestinal lymph changes after intragastric administration of single doses of diets containing either 3 g per kg ethanol (●——●) or isocaloric carbohydrate (O-----O) to eight pairs of male rat litter mates pair-fed liquid diets containing either ethanol or additional carbohydrate for three to four weeks. The rats received in randomized sequence either alcohol-containing or control diet at six-hour intervals. Dietary lipids were labelled with [carboxy-^{14}C] tripalmitin. Results are means and standard error of means. (Baraona & Lieber, 1975.)

observations but enhanced lymph flow induced by neostigmine also increases protein flow and augments the transport of dietary lipid. Its effect is probably due to alterations in intestinal blood flow. Chronic ethanol administration appears to inhibit the acute effects of ethanol on lymph flow and lipid transport. Again, there is no good explanation for this observation at present.

The transport of lipids in intestinal lymph

Circulating lipids in blood and lymph are chiefly transported as lipoproteins, though some lipids, such as free fatty acids, are bound to circulating albumin. Classification of human plasma lipoproteins is shown in Table 7.1. The most common classification, that is according to the density of the lipoproteins, defines four main groups: chylomicra, very low density lipoproteins (VLDL), low density lipoproteins (LDL) and high density lipoproteins (HDL). A group of very high density lipoproteins is also recognised (VHDL). There is evidence that all four main classes of lipoprotein are synthesised in the intestine, but for VLDL, LDL and HDL the liver is the major site of synthesis (Schumaker & Adams, 1969).

This section considers the two types of lipoprotein which are responsible for the major amount of fat transport from the intestine via the mesenteric lymph, namely chylomicra and VLDL. The intestinal mucosa is known to synthesise lipoproteins of density less than 1.006. The intestine is a major source of these lipoproteins in lymph and intestinal microsomes can synthesise the β-apoprotein of these lipoproteins (Kessler, Stein, Dannacker & Narcessian, 1970), although other lipoprotein apoproteins are also synthesised by the intestine. Some smaller peptide components of intestinal lymph VLDL are probably derived from plasma lipoproteins by filtration in the gut (Windmueller, Herbert & Levy, 1973).

Chylomicra

Chylomicra are particles of lipid present in large numbers in the intestinal lymph during fat absorption, and a large proportion of dietary lipid absorbed by the small intestine is taken to the systemic circulation in these triglyceride-rich particles. The term 'chylomicron' was coined in 1920 by Gage to refer to microscopic bodies seen in post-prandial plasma, and in this discussion the term has been exclusively applied to those low density lipid particles which arise from intestinal mucosal epithelium. Since 1920, methods for quantifying dimensions and composition of such lipid particles have been developed and applied to a broad spectrum of lipid particles. As a result,

TABLE 7.1. *Composition and properties of human plasma lipoproteins.* (Jackson, Morrisett & Gotto, 1976.)

Properties	Chylomicra	VLDL	LDL	HDL	VHDL
Density (g per ml)	< 0.95	0.95–1.006	1.006–1.063	1.063–1.210	1.210–1.250
Electrophoretic mobility	Origin	Pre-β	β	α	α
Major apoproteins	ApoB ApoC-I ApoC-II ApoC-III	ApoB ApoC-I ApoC-II ApoC-III Arginine-rich protein	ApoB	ApoA-I ApoA-II	ApoA-I ApoA-II
Minor apoproteins	ApoA-I ApoA-II	Thin-line protein ApoA-I ApoA-II		ApoC-I ApoC-II ApoC-III Thin-line protein Arginine-rich protein	

the term chylomicron has more recently been reserved for those particles having a flotation constant (Sf) greater than 400. This criterion divides chylomicra from a group of very low density lipoproteins (VLDL) of smaller particle size. The particle size and flotation characteristics of VLDL, however, merge with those of the chylomicra. The very high proportion of triglycerides in chylomicra is responsible for their low density.

Chylomicra which have been examined microscopically after embedding and sectioning have a circular outline (Jones, Thomas & Scott, 1962; Kay & Robinson, 1962; Jones *et al.*, 1963; Schoefl, 1968) and electron micrographs of chylomicra prepared by shadowing show them to be spherical (Bierman *et al.*, 1966). From measurements using dark field microscopy, the diameter of chylomicra was originally taken to be between 0.5 and 1.0 μm. Pinter & Zilversmit (1962) examined the particle size by sucrose density gradient centrifugation. By this method, the mean particle diameter of dog chylomicra has been found to be between 0.18 and 0.29 μm with few particles greater than 0.5 μm. Similar data were obtained for rat chylomicra (Zilversmit, Sisco & Yokoyama, 1966).

Fraser (1970) has studied the particle size of three groups of chylomicra obtained from the thoracic duct lymph of rabbits fed with corn oil. After separation of the groups by density gradient centrifugation, the dimensions of the particles as measured in electron micrographs were compared with the theoretical values obtained from their flotation. Theoretical diameters for particles of Sf greater than 10000, 1000–10000 and 400–1000 are approximately 360 nm, 120–360 nm and 75–120 nm respectively. In the last two categories the measured particle sizes corresponded quite well with theoretical values, but in the group of Sf greater than 10000 there were many particles of diameter less than the expected 360 nm. It is possible that anomalous behaviour of some groups of chylomicra may have occurred during density gradient centrifugation perhaps as a result of aggregation and deaggregation of particles during the process of preparation. The size of chylomicra is influenced by the dietary fat load. Fraser *et al.* (1968), using electron microscopy, found in rabbits that a high fat diet leads to a marked increase in the diameter of the chylomicra, and Bouquillon, Carlier & Clement (1974) have obtained similar results in rats. Chylomicron size has been shown by various investigators to be related to the composition of the dietary fat. For example, in the rabbit, chylomicra derived from peanut and corn oils, which contain relatively large amounts of linoleic acid, have somewhat more phospholipid and are smaller than those derived from coconut oil or an emulsion of oleic acid and glyceryl mono-oleate (Redgrave & Dunne, 1975). In this study median diameters of chylomicra were calculated to range from 104 to 152 nm.

On chemical analysis, chylomicra prepared from lymph are found to consist of large amounts of triglyceride (85–90%), together with phospholipids (6–9%), free and esterified cholesterol (3%) and protein (0.5–2%) (Lindgren & Nicholls, 1960). Of the phospholipids of chylomicra, the principal species is lecithin (phosphatidyl choline) (75%), but small amounts of phosphatidyl ethanolamine (12%), sphingomyelin (5%) and lysolecithin (3%) are also present (Zilversmit, 1965). Since the bulk of chylomicron lipid is non-polar triglyceride, a comprehensible structure for these spherical bodies would comprise an oil droplet, consisting of liquid triglyceride containing the relatively small amounts of esterified cholesterol dissolved in this medium, and a stabilising layer of the more polar lipid (the phospholipid) together with protein lying at the interface between the oil and the aqueous medium. Cholesterol could be considered as partitioned between the oil interior and the more polar surface layer; in the latter situation, it is possibly complexed with the surface protein.

High resolution electron micrographs of washed chylomicra show that there is a coat on the surface of these bodies which has an irregular structure and is more osmiophilic than the contents (Salpeter & Zilversmit, 1968). The appearance of this coat is not at all similar to the plasma membranes of cells or to intracellular membranous structures, but is compatible with the presence of a polar lipoprotein or lipid monolayer on the surface of the lipid droplet.

It has been possible by two techniques, repeated freezing and thawing or dehydration on a rotary evaporator followed by hydration and dehydration, to prepare fractions of chylomicra from dog, rat and man which represent the oil core of the particles from a 'membrane' fraction. This latter fraction, which is believed to represent the coat or 'ghost' of the chylomicron, can be recovered as a pellet on high speed centrifugation (Zilversmit, 1965). Plate 7.2 shows an electron micrograph of this fraction. These 'membranes' consist of protein, together with all the chylomicron phospholipid, small amounts of triglyceride, and some free cholesterol. The oil phase contains all the esterified cholesterol, most of the triglyceride, 25–35% of the free cholesterol, but none of the phospholipid.

Thus the chemical structure of the chylomicron appears in general to conform with the theoretical model. In keeping with this, too, is the relationship which has been demonstrated between chylomicron size and chemical composition. Using density gradient centrifugation, Yokoyama & Zilversmit (1965) prepared groups of chylomicra of various sizes and found that as the particle size fell, so did the proportion of triglyceride in the particle, while the proportion of the surface components, protein and phospholipid, rose. Fraser *et al.* (1968) and Fraser (1970) have shown a close correlation between the volume : surface area and the triglyceride content : phospholipid content ratios of chylomicra. Such an observation is consistent with the hypothesis that a layer of phospholipid coats the surface of a droplet of less polar lipid. These authors have also calculated from their data that the mean area of the chylomicron surface occupied by a single phospholipid molecule is approximately 0.6 nm².

Of the substances associated with the outer coat of chylomicra, i.e. the phospholipids, protein and cholesterol, the phospholipid probably plays the major part in stabilising the chylomicron. Phospholipase treatment of chylomicra destabilises the droplets while treatment with proteolytic enzymes does not have this effect (Elkes & Frazer, 1943; Robinson, 1955).

Transport of absorbed lipids

Discharge of chylomicra from the cell

From what has been said already, it is apparent that in the epithelial cells droplets of lipid are seen to accumulate within membrane-bound compartments of the cell during fat absorption. Cardell *et al.* (1967) noticed that vesicles, apparently derived from smooth endoplasmic reticulum, containing fat droplets are present in considerable numbers in the cytoplasm close to the lateral borders of the cell during fat absorption, and also at this time, enlarged intercellular spaces appear between neighbouring epithelial cells at the level of their nuclei, that is towards the basal poles of the cells. These intercellular spaces contain numerous lipid droplets of 0.05–0.5 μm diameter. These droplets lack the well-defined membrane which surrounds their counterparts within the cell. Palay & Karlin (1959b) proposed that an extrusion process, a reversed pinocytosis, delivered these chylomicra to the intercellular space. This proposed mechanism is generally accepted, but stages in the process are hard to demonstrate in morphological studies and it is possible, therefore, that this is a rather rapid phenomenon. From the extracellular space, chylomicra appear to pass through the basement membrane of the epithelial cells into the lamina propria of the villus and thence gain access to the lymphatic vessel, chiefly by passage between lymphatic endothelial cells (see Plate 3.2). The means by which chylomicra make their orderly way from the epithelial cell through the lamina propria to the lymphatic vessel are not understood.

The mechanisms by which lipoprotein particles are discharged from the enterocyte through its lateral cell wall by reversed pinocytosis are unknown. Recent studies by Glickman, Perotto & Kirsch (1976) have attempted to examine the question of whether microtubules may be involved in this process. There is some reason to believe that this might be so since microtubules are well known to have important effects in cellular function, including spindle formation during mitosis, and some types of cell secretion; the latter include secretion of noradrenaline, collagen, insulin and very low density lipoproteins (VLDL) formed in the liver. Microtubules are believed to have a directional function in cell secretion, that is, they align secretory granules close to the plasmalemma.

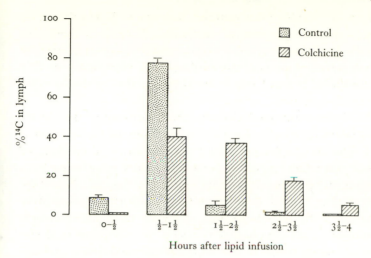

Fig. 7.6. Effect of colchicine on the lymphatic transport of [^{14}C]oleic acid. Colchicine (0.5 mg per 100 g) was administered intraperitoneally two hours before the intra-duodenal infusion of a micellar solution of [^{14}C]oleic acid. Lymph radioactivity was determined in lymph samples from control and colchicine-treated animals. Results are means plus standard error of means for five rats. (Glickman *et al.*, 1976.)

When colchicine, which is well-known for its ability to disrupt microtubular structure and inhibit re-polarization of microtubular subunits, was given intraperitoneally in a dose of 0.5 mg per 100 g body weight to rats with indwelling mesen-teric lymph cannulae it caused a marked delay in the lymph-atic absorption of oleic acid (Fig. 7.6). No difference was noted in the apoprotein composition of the lipoproteins of test versus control animals but it was apparent that colchicine treatment was associated with a holdup of lipid in the mucosal cells. Microscopic examination of these cells suggested that most of this residual lipid was contained within the endoplas-mic reticulum and Golgi saccules as particles comparable in size with chylomicra. The results were taken to indicate an impairment in lipid transport at a late stage in its transcellular movement, affecting the exit of these lipid particles from the cell and possibly involving interference with microtubular function. It is difficult to be certain that microtubules are involved in the process since the evidence obtained with the use of colchicine is open to the criticism that this drug is known to affect not only microtubule formation but other cell functions including protein synthesis. (For a review, see Stephens & Edds, 1976.) Further studies are required to establish whether or not microtubules do participate in this important step in lipid transport through the epithelial cells of the small intestine.

As has been mentioned already, the outer coat of the chylomicron bears no morphological resemblance to plasma membrane. The oil droplet produced within intestinal epithelium does not, therefore, acquire a coat of plasma membrane by a 'pinching off' process during its exit from the cell. Nor does it appear that the oil droplet collects its surface lipids after leaving the cell, by adsorption of lipids from other lymph lipoproteins. This conclusion is based on comparison of the fatty acid composition of chylomicron phospholipid and the phospholipid of the remaining subnatant fluid of lymph, obtained by centrifugation. The subnatant contains lipoproteins of higher density (Table 7.2). Chylomicron lecithin contains a much higher percentage of linoleic acid than the lecithin of lymph subnatant which has, on the other hand, a much greater concentration of arachidonic acid than chylomicron lecithin. The palmitic acid content of chylomicron sphingomyelin is much greater than that of the lymph subnatant sphingomyelin. These and other similar data lead to the conclusion that chylomicron phospholipid is not merely acquired by casual 'adsorption' of the phospholipids of lymph but must be added to the oil droplet by a specific intracellular process, as is the case for the protein component of chylomicra (Zilversmit, 1968).

The importance of apoprotein synthesis for chylomicron secretion has already been discussed. Under certain circumstances, factors other than the synthesis of apoproteins may limit the rate of chylomicron formation and discharge from the enterocyte. When triglyceride is infused into the small intestine of rats at high rates it is clear that the distal small intestine accumulates triglyceride in the enterocyte to a greater extent than the proximal small intestine (Sabesin, Holt & Clark, 1975). These rats received no inhibitors of protein synthesis. This suggests a basic difference in the capacity of these cells for chylomicron formation though whether this is a qualitative or

TABLE 7.2. *Fatty acid composition of phospholipids, as % of total methyl esters.* (From Zilversmit, 1968.)

Lymph infranatant was prepared by centrifugation at 64000–75000 g for 60–90 min to remove chylomicra.

a Phosphatidyl choline

	16:0	18:0	18:1	18:2	20:4
			Fatty acid (no. carbon atoms: unsaturated bonds)		
Lymph chylomicra ($n=5$)	15.6	24	11.8	41.8	4.1
Lymph subnatant ($n=4$)	14.25	26.5	10.6	25.5	17.25

b Sphingomyelin

	16:0	18:0	18:1	18:2	20:0	22:0	23:0	24:0	24:1
Lymph chylomicra ($n=4$)	56.75	12.5	2.4	1.8	4.8	3.3	3.5	3.0	5
Lymph subnatant ($n=2$)	24	13	1.1	1.2	4.4	6.5	9.6	8.8	18

quantitative difference is unclear. It would be interesting to determine whether there would be any functional adaptation of the lipid absorption of ileal enterocytes in response to jejunectomy. Sabesin *et al.* (1975) suggest that other factors may be rate-limiting and one candidate may be the formation of the phospholipid component of chylomicra. Thus O'Doherty, Kakis & Kuksis (1973) demonstrated impaired fat release from the intestinal mucosa of rats deprived of dietary or biliary lecithin and of choline. However, absence of phospholipid or its precursors in the gut lumen led also to depressed apoprotein synthesis in the mucosa so these data are hard to interpret. At present, the importance of phospholipids relative to apoproteins in chylomicron secretion must be considered unsettled.

The uptake of chylomicra by lymphatic vessels

During the post-prandial period, a large movement of chylomicra takes place from intestinal epithelial cells to the lamina propria of the villi and thence into the central lacteal. The basement membrane is fragmentary at best and does not impede chylomicron uptake into the lacteal. The mode of uptake has been studied by several groups, using morphological methods (see also Chapters 2 and 3). Using techniques which demonstrate both chylomicra and lipoproteins in electron microscopic preparations Casley-Smith (1962) demonstrated that both these particles enter lymphatics of the rat jejunum by two distinct routes, either through the endothelial cells or between them. He considers the latter to be quantitatively more important.

This view has been challenged by Dobbins (1971) who believes the transcellular route through membrane-bound vesicles to be of prime importance for molecules of molecular weight greater than 40000. Dobbins & Rollins (1970) used peroxidase (MW 40000; diameter 5 nm) and chylomicra (diameter 100 nm–1 μm) as tracers. These particles could be shown traversing the endothelial cell in vesicles, though it should be noted that while it is relatively easy to show foreign particles passing into or through the cell in this way it is difficult, for reasons not yet clear, to demonstrate transcellular transport of chylomicra.

The energy requirements for this pinocytotic transport are not known (Bruns & Palade, 1968) nor is the reason for the apparent polarity of the mechanism such that chylomicra are transported *into* the lacteal. Dobbins (1971) suggests that the mechanism operates in both directions but that movement of lymph creates a concentration gradient of chylomicra from villus interstitium to lacteal and that chylomicra effectively 'diffuse' down this concentration gradient.

Modification of chylomicron composition in lymph

It is clear from preceding discussion that the composition of the phospholipids of the surface of chylomicra is different from that of phospholipids of other lipoproteins in the subnatant fraction, obtained by centrifugation of lymph. It can therefore be concluded that *rapid* equilibration of chylomicron phospholipid with phospholipid of clear lymph does not occur. The possibility of a slow exchange of chylomicron surface lipids with those in lymph or plasma is, however, not excluded, and studies by Minari & Zilversmit (1963) indicate that both molecular exchange and alterations in the total amounts of lipid species in chylomicra could occur in lymph or plasma. In a study of dog lymph chylomicra incubated with serum *in vitro*, these workers found a progressive accumulation of free cholesterol in the particles and a fall in phospholipids, chiefly lecithin. The lecithin lost from the chylomicra appeared mainly in high density lipoproteins of the serum. By contrast, little change was observed in the chylomicron content of triglyceride, cholesterol ester or sphingomyelin during the incubation. In addition to the net changes in chylomicron phospholipid and cholesterol, it was possible to demonstrate exchange of molecules of chylomicron phospholipid with those of serum by measuring the change in specific activity of chylomicron phospholipid labelled with ^{32}P.

These results clearly show that modification of surface lipids of chylomicra can occur during their life in serum and, by inference, substantial alterations in chylomicron composition may occur during the transit of these particles through the lymphatic system. The modifications in chylomicron lipid in the general circulation are probably mainly due to the action of circulating lipoprotein lipase and to lecithin–cholesterol acyl transferase (Schumaker & Adams, 1969). The extent to which these enzymes modify chylomicron components in lymph is not known. A recent study by Clark & Norum (1977) has raised new interest in this question. They have shown that although lecithin–cholesterol acyl transferase is present in the intestinal lymph of fasting rats its concentration is less than 10% that in serum; however, following feeding its concentration in lymph rises substantially, possibly due to an input of the enzyme by the gut.

Very low density lipoproteins (*VLDL*)

This class of triglyceride-rich particles, of Sf values 20–400, also appears to transport considerable amounts of lipid from the gut via the lymph to the general circulation. In *fasting* rats, Ockner, Hughes & Isselbacher (1969a) found both VLDL and chylomicra in mesenteric lymph, and under these circumstances VLDL was the major lipoprotein species of lymph. It was found to contain 47% of the triglyceride and about 54% of the cholesterol of lymph, while chylomicra carried only 40% of the triglyceride and 17% of the cholesterol. The shift towards a predominance of chylomicron lipid in lymph occurs on fat feeding. By interfering with lipid absorption from the small intestinal lumen either by bile diversion or micellar disruption with cholestyramine, it was shown that approximately 80% of the total intestinal lymph triglyceride and cholesterol, and essentially *all* that of VLDL and chylomicra comes from the intestinal lumen. Using immunoelectrophoresis, these workers have shown that lymph and plasma VLDL share a common major apoprotein. Certain qualitative differences exist in the fatty acid composition of lymph and plasma VLDL. It is therefore possible that lymph VLDL derived from intestinal mucosa makes a significant contribution to total plasma VLDL and the lipid composition differences might be explicable on the basis of addition of triglyceride of hepatic origin to plasma VLDL.

The mechanism by which glycerides resynthesised in the intestinal mucosa are assigned to carrier lipoproteins appears to be rather specific (Ockner *et al.*, 1969b). In rats with intestinal lymph fistulae, feeding palmitic, oleic or linoleic acid leads to an increase in chylomicron triglyceride. In addition, palmitic acid feeding results in a two-fold increase in VLDL triglyceride, whereas the absorption of oleic or linoleic acid does not significantly change the amounts of VLDL triglyceride. The cholesterol distribution between chylomicra and VLDL is also influenced by the species of fatty acid being absorbed, since during palmitic acid absorption VLDL cholesterol is approximately equal to chylomicron cholesterol, but during the absorption of the unsaturated fatty acids the total chylomicron cholesterol is more than twice that of VLDL. In sheep, absorbed long-chain fatty acids are mainly transported as VLDL rather than chylomicra and in this species most palmitic, stearic and oleic acids are transported as triglyceride whereas essential fatty acids, linoleic and linolenic are transported to a considerable extent in lymph phospholipids (Leat & Harrison, 1974).

Lymphatic transport of lipids in the fasting state
Most lipid species in the diet are diluted to some extent in the intestinal lumen by an endogenous pool of lipid derived from bile and shed mucosal epithelium. It is also possible that chylomicra and other fat particles may be exuded across the mucosa into the intestinal lumen and contribute to this pool. Dietary triglycerides and fatty acids enter such an endogenous pool; a single balanced meal would be expected to swamp the endogenous pools of these species of lipid. Fat feeding causes a rise in the concentration and transport of chylomicra in the lymph and this is sustained over several hours. However, in the fasting animal, lipids associated with very low density lipoproteins (VLDL), particles which are smaller than chylomicra, are being constantly transported in intestinal lymph (Ockner *et al.*, 1969a). Mattson & Volpenhein (1964) found fat equivalent to approximately 150 mg of oleic acid per day transported in the thoracic duct lymph of rats fed a fat-free diet. In addition to endogenous lipid absorbed from the intraluminal pool, mucosal synthesis of triglyceride, phospholipid and sterols contributes to the transport of endogenous lipid in intestinal lymph in the fasting state. Fat feeding results in an

increase in the transport of endogenous fat in lymph (Karmen, Whyte & Goodman, 1963). Shrivastava, Redgrave & Simmonds (1967) demonstrated that the output of esterified fatty acid in the thoracic duct lymph of fasting rats was decreased by 80% when a bile fistula was produced. This could not be corrected by infusion of sodium taurocholate at 5 or 10 mg per hour, and it was shown that the loss of biliary lipid could quantitatively account for the fall in lymph lipid observed on bile diversion. It was found by gas–liquid chromatographic analysis of biliary, lymph and plasma lipids, that before biliary drainage was established the lymph lipid fatty acids were similar to those of bile but after bile diversion they took on the qualitative pattern of plasma lipids. These authors suggested that fatty acids of biliary phospholipid contribute a major amount to fasting lymph lipid, and that after bile diversion the much smaller amounts of lipid transported in lymph are mainly derived from lipoproteins filtered from plasma.

Intraduodenal ethanol enhances intestinal mucosal triglyceride synthesis and the production of non-dietary VLDL by rat intestine (Mistilis & Ockner, 1972). Although this lipid production in itself is unlikely to explain entirely the accumulation of lipid in the liver of these animals which occurs in response to intraduodenal ethanol, it undoubtedly contributes to overall lipid balance in the animal, since lymph diversion partly prevents hepatic triglyceride accumulation (Ockner, Mistilis, Poppenhausen & Stiehl, 1973).

The partition of fatty acids between the portal venous blood and intestinal lymph

It is generally accepted that long-chain fatty acids of fourteen or more carbon atoms are transported mainly as resynthesised triglyceride in the chylomicra of lymph, and that fatty acids of ten carbon atoms or less are carried primarily as free fatty acids in the portal vein. Fatty acids of intermediate chain length are thought to be absorbed by both routes. This general scheme has emerged from a large series of studies; as will be seen, there are a number of qualifications to be made to this overall principle.

Studies by Bloom *et al.* (1950), Bloom, Chaikoff & Reinhardt (1951), Chaikoff *et al.* (1951), Reiser & Bryson (1951), Bergström, Blomstrand & Borgström (1954) and Blomstrand (1954) have all indicated that long-chain fatty acids presented as either the free acid or as esters of glycerol are transported to the circulation after intestinal absorption primarily in lymph chylomicra. However, in many of these studies there are indications that even under normal circumstances small amounts of long-chain fatty acids may gain access to the general circulation by routes other than the lymphatic one. To take an early investigation in detail, Bloom and his colleagues (1950) studied the uptake and transport of [^{14}C]palmitate, given either as the free acid or as tripalmitin to unanaesthetised rats with thoracic duct or intestinal lymph fistulae. In nineteen to twenty-four hours, twelve of the fourteen rats studied had absorbed about 90% of the ^{14}C of the administered fatty acid; 70 to 92% of this was recovered in lymph lipid of animals with thoracic duct fistulae and 69 to 84% in lymph lipid of rats with intestinal lymphatic fistulae. There remained therefore a small amount of label unaccounted for, and it was questioned whether portal venous transport accounted for a minor proportion of absorbed long-chain fatty acid. Blomstrand (1954) found that the major part of absorbed ^{14}C-labelled linoleic acid is transported by the thoracic duct as triglyceride, but he observed that some labelled carbon dioxide was expired, despite an external thoracic duct fistula, and again a minor participation of the portal venous system in the transport of long-chain fatty acids was postulated. This experiment is, of course, subject to the objection that lymphatico-venous communications in the mesenteric region might account for the observations.

As will be seen, failure of esterification of absorbed fatty acids within the mucosal cell re-routes long-chain fatty acids into the portal venous blood to some extent. Before considering abnormal situations where this occurs extensively, it is appropriate to ask to what extent this portal transport of long-chain fatty acids ordinarily occurs. There is no easy answer to this problem since most of the studies have been carried out in rats and apparently the handling of fatty acids by intestinal mucosa differs among rat strains. In hooded rats, mucosal esterification is less extensive and a higher proportion of fatty acids travels by the portal vein than in albino rats where more extensive triglyceride resynthesis occurs (Johnston, 1968).

Short-chain fatty acids are transported by the portal venous route to a large degree. Blomstrand (1955) found, on feeding rats with $[^{14}C]$decanoic acid (C_{10}), that only a minor proportion (3–16%) of the fatty acid was transported by thoracic duct lymph despite almost complete absorption of the label over twenty-four hours. Such ^{14}C as was present in the lymph was esterified as triglycerides and phospholipids, and there was a rapid appearance of large amounts of ^{14}C in expired carbon dioxide, suggesting extensive portal venous transport of this fatty acid and rapid metabolism. Borgström (1955) was able to show in rats fed this same labelled fatty acid that the concentration of lipid ^{14}C in portal venous blood was two to four times greater than that in inferior vena cava blood and that in the portal venous blood 64–74% of the ^{14}C was in the form of unesterified decanoic acid, presumably bound to protein. Earlier studies by Bloom *et al.* (1951) with labelled lauric (C_{12}) and decanoic acid (C_{10}) showed that of the absorbed material only 15–55% in the case of lauric and 5–19% in the case of decanoic acid could be recovered in lymph, while with labelled stearic (C_{18}) and myristic (C_{14}) acids nearly all the absorbed label was recovered in the lymph.

More recent work has re-examined the partition of fatty acids between the portal venous and mesenteric lymphatic routes. Hyun, Vahouny & Treadwell (1967) used rats with thoracic duct fistulae and portal venous cannulae. The fatty acids studied, ^{14}C labelled octanoic acid (C_8), oleic acid (C_{18}) and 2-ethyl-n-hexanoic acid (C_8), were given intragastrically. It was found that 85% of absorbed oleic acid was transported by the lymphatic system but that as much as 15% was transported directly by portal venous blood. With the shorter-chain fatty acids, about 95% of the absorbed acids was transported by the portal system. Incidentally, the straight-chain octanoic acid was found to be more rapidly absorbed than its branched-chain analogue, 2-ethyl-n-hexanoic acid. When lymph lipids were studied, 85% of the lymph radioactivity after oleic acid administration was shown to be present as triglyceride, but all of the shorter-chain fatty acid radioactivity in lymph was present as free fatty acid. In the portal vein, 50% of the radioactivity was present as free fatty acids when labelled oleic acid was given, whereas 100% of the radioactivity was in the form of free fatty acids when the labelled shorter-chain fatty acids were given.

Kayden & Medick (1969) have approached the fatty acid partition problem in an interesting way. In their experiments, rats were treated with puromycin, an inhibitor of protein synthesis which is thought by some workers to produce an inhibition of chylomicron formation, and hence inhibition of long-chain fatty acid transport, by suppressing intestinal mucosal synthesis of β-lipoprotein (see the section on mucosal protein synthesis and fat transport, p. 171). As judged by the excretion of ^{14}C-labelled carbon dioxide in expired air, these rats absorbed labelled octanoic acid as efficiently as control rats but the transport of long-chain fatty acids in chylomicra was inhibited in the puromycin-treated animals. The lack of effect of puromycin on the absorption of octanoic acid and trioctanoin is in keeping with the hypothesis that these fats are chiefly transported as albumin-bound unesterified fatty acids in the portal vein. However, in both the puromycin-treated and control animals fed labelled long-chain fats, i.e. oleic acid or triolein, the amounts of long-chain fatty acid converted to carbon dioxide were the same, which suggests that a large amount of long-chain fatty acid must have entered the circulation in a form other than as chylomicra transported in lymph. Rats were then prepared with thoracic duct fistulae and divided into puromycin-treated and control groups. An inverse relationship was found in those animals treated with puromycin, between the radioactivity in lymph and the radioactivity in expired carbon dioxide. In other words, as puromycin reduced the lymphatic transport of labelled long-chain fatty acid, so an alternative route must have delivered the fatty acid to the circulation. It was concluded that long-chain fatty acid which could be released both by digestion of glycerides within the small intestinal lumen and by digestion of absorbed mono-glycerides by a lipase in intestinal mucosa (Senior & Isselbacher, 1963) could be delivered in substantial amounts to the circulation by being initially transported in portal venous blood, bound to albumin. These results when interpreted in the light of a straightforward situation, i.e. an inhibition of β-lipoprotein synthesis by mucosal cells under the influence of puromycin, are in keeping with the general hypothesis of fatty acid partition between lymph and portal blood. There is, however, some doubt about the nature of the experimental model created by giving rats puromycin (see section on mucosal protein synthesis and chylomicron formation p. 171), and these results of Kayden & Medick have to be interpreted in the light of a much more complex situation which creates defects in fat absorption. Nevertheless, their results indicate that disorders of mucosal handling of lipids may result in diversion of considerable amounts of long-chain fatty acid towards the portal venous route.

Another situation which undoubtedly results in a substantial increase in the transport of long-chain fatty acids by the portal vein is bile salt deficiency. Borgström (1953), giving [^{14}C]palmitic acid to rats with bile fistulae, found that only about one-quarter of the absorbed radioactivity appeared in intestinal lymph, and he suggested that bile salt deficiency would interfere not only with the intraluminal phase of fat digestion and absorption, but also with mucosal metabolism of the absorbed fatty acids, perhaps re-routing them from the lymph to the portal vein. This hypothesis implies that some bile salts penetrate the absorbing epithelium of the upper small intestine. Since then, this hypothesis has gained further support: Dawson & Isselbacher (1960) demonstrated that con-jugated bile salts stimulate the esterification of palmitic acid in slices of rat small intestine *in vitro*, and Saunders & Dawson (1963) found that when ^{14}C-labelled oleic acid was fed to rats with bile fistulae, approximately half the radioactivity was absorbed but only 7.2% of this was transported in the lymph, and an abnormally high proportion (28%) of the lymph lipid radioactivity was present as free fatty acid. (Normal values for free fatty acid as a percentage of lymph lipid during fat absorption have been quoted as 3.8–11.8% by Borgström & Tryding, 1956.) Taurocholate replacement altered the propor-tion of lymph free fatty acid towards normal. It was also noted that there were considerable amounts of radioactivity in the portal vein, and it was concluded that bile salt deficiency reduced mucosal esterification of fatty acid and directed a greater proportion of fatty acid than usual into the portal vein. Gallagher, Webb & Dawson (1965), have further studied the effect of bile exclusion on the absorption of [14C]oleic acid and [14C]triolein in the rat. Despite bile diversion, substantial amounts of lipid were absorbed but a large proportion of the absorbed fat in the lymph was in the free fatty acid fraction. Morgan and Borgström (1969) showed that during the assimilation of labelled triglyceride in rats with an intact biliary system or a bile fistula the proportion of absorbed fatty acid to monoglyceride found in the lymph changed in the bile-diverted animals such that fatty acids appeared to be absorbed in excess of monoglyceride. In the absence of a micellar solution it appears that fatty acid absorption is less seriously impaired than monoglyceride absorption though the explanation for this is at present unclear.

It is interesting that bile salt deficiency appears to direct other lipid species, which are normally believed to be almost exclusively transported by lymph, towards the portal vein. Vitamin A is an example of such a lipid which may be transported in the portal vein to some extent in the bile fistulae animal (Gagnon & Dawson, 1968). It may be that a failure of mucosal esterification of the vitamin occurs in the absence of bile salts.

Thus it seems from the foregoing evidence that the portal venous transport of long-chain fatty acids can occur in normal animals to a limited extent (a mean estimate being approximately 10%) and to a much greater extent where mucosal handling of absorbed lipid is abnormal, i.e. in bile fistula and possibly in puromycin-treated animals. Another experimental situation which has been considered to produce this effect is the feeding of 2-ethyl-n-hexanoic acid to rats. This branched-chain analogue of octanoic acid is thought to interfere with the re-esterification of long-chain fatty acids in the mucosal cell and might channel a greater proportion of such fatty acids in their free form into the portal vein (Hyun *et al.*, 1967). Human a-β-lipoproteinaemia may be a condition where a substantial amount of long-chain fatty acid is transported by the portal vein

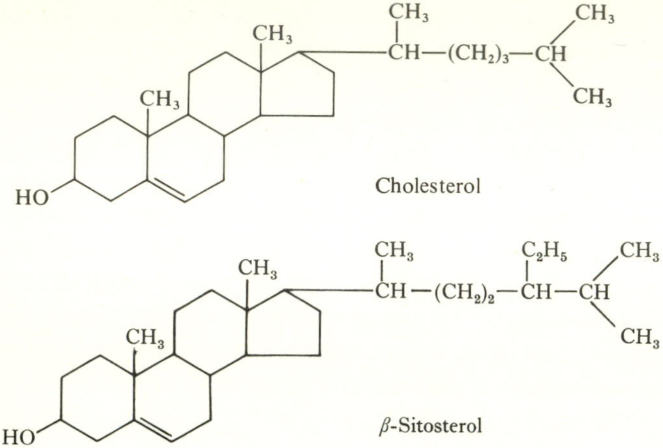

Fig. 7.7. Structure of cholesterol and β-sitosterol.

thereby ameliorating the steatorrhoea from which these patients suffer (see above).

Recently, Jodal (1973) has suggested that the existence of a counter-current exchange mechanism in the villi of the small intestinal mucosa may delay the absorption of long-chain fatty acid allowing time for re-esterification and thus determining the partition of these fatty acids in favour of lymph. This hypothesis requires further experimental study.

Absorption and transport of minor dietary lipids

Sterols
Sterols and their fatty acid esters are important constituents of many tissues. They are virtually insoluble in water and the major sterol found in animal tissues, cholesterol, plays a particularly important part as a structural component of cell membranes. Another vital aspect of cholesterol as a biological substance is its role as precursor for the formation of bile salts and steroid hormones. Figure 7.7 shows the structures of cholesterol and β-sitosterol, a closely related plant sterol. The other major plant sterols are campesterol and stigmasterol.

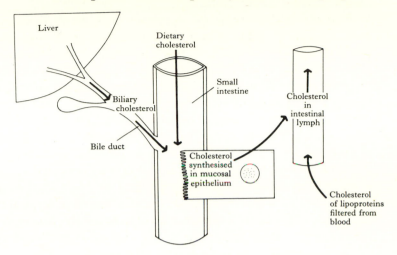

Fig. 7.8. A schematic representation of the sources of lymph cholesterol.

Overall scheme of cholesterol absorption

Dietary (exogenous) cholesterol, which is estimated to be 0.5–2 g per day in Western diets, is derived from foods of animal origin and mixes in the lumen of the small intestine with endogenous cholesterol from bile and from shed mucosal epithelium (Fig. 7.8). That the small intestinal mucosa can synthesise cholesterol has been demonstrated by Dietschy & Siperstein (1965) and this cholesterol makes a substantial contribution to the serum cholesterol (Lindsey & Wilson, 1965). The presence of considerable amounts of cholesterol in bile greatly complicates any study of absorption of exogenous cholesterol since fat feeding, by promoting bile flow into the small intestine, will augment the amounts of endogenous cholesterol presented to the mucosa for absorption. Similar considerations apply to phospholipid absorption.

Cholesterol exists in three main physico-chemical states in the intestine – in solution in any oil phase present, in bile acid micelles and as a crystalline precipitate; some cholesterol has also been shown to form a complex with intestinal mucoproteins and it may also absorb to dietary fibre. Cholesterol can only be taken up into mucosal epithelium from the micellar phase (Treadwell & Vahouny, 1968) and complete bile diversion abolishes cholesterol absorption. It is probable that cholesterol in precipitate in intestinal content is in equilibrium with micellar cholesterol and is thus available for absorption. Hydrolysis of neutral lipid by pancreatic lipase enhances micellar solubilisation of cholesterol by expanding the bile acid micelles with amphiphiles (monoglycerides and fatty acids) (Hofmann & Small, 1967). Cholesterol is held in such a mixed micellar solution by non-polar solubilisation in the hydrophobic centre of the micelle. This presumably accounts for the enhancement of cholesterol absorption by concomitant fat feeding (Treadwell & Vahouny, 1968).

In addition to the free sterol, the diet contains some fatty acyl esters of cholesterol. Hydrolysis of these esters, a necessary preliminary to absorption of the sterol (Vahouny & Treadwell, 1964), is catalysed by a pancreatic esterase. This enzyme requires that its substrate be in micellar solution. It is interesting that tri-hydroxy bile acids are able to protect this enzyme from proteolytic inactivation by other pancreatic hydrolases (Treadwell & Vahouny, 1968). Although hydrolysis of sterol esters is a requisite for absorption of the sterol, cholesteryl ethers of aliphatic alcohols can be absorbed intact in rats, and are recovered in thoracic duct lymph. The longer the aliphatic chain the less the degree of absorption (Borgström, 1968*a*).

Cholesterol absorption is known to be incomplete (Treadwell & Vahouny, 1968) whereas the fatty acids of ingested long-chain triglycerides are absorbed quantitatively (Bloom *et al.*, 1951; Borgström, Dahlqvist, Lundh & Sjövall, 1957). In the rat it appears that an almost constant fraction of exogenous cholesterol (approximately 40%) can be recovered in thoracic duct lymph after the administration of different amounts of cholesterol dissolved in glyceryl trioleate.

Cholesterol is absorbed primarily in the upper small intestine (Arnesjö, Nilsson, Barrowman & Borgström, 1969). From mixed micelles the transfer of the sterol into the mucosal epithelium is probably a passive process in which the sterol initially dissolves in the lipid membrane of the cell while the major amount of micellar bile salt remains in the lumen of the intestine. At this uptake stage, considerable specificity of sterol transport is evident since cholesterol is taken up approximately ten times more readily than β-sitosterol (Borgström, 1968*b*).

This discrimination between such structurally similar compounds (see Fig. 7.7) by the cell membrane is destroyed by metabolic inhibitors which might in some way alter the configuration of the membrane (Sylven, 1970).

The transfer of cholesterol through the cell to its inclusion as a component of chylomicra and of VLDL is poorly understood, though it must be presumed that it is held in solution in various lipid phases or bound to protein in its transit through the cell. Absorbed cholesterol mixes with newly synthesised mucosal cholesterol in the cell and a large proportion (approximately 70%) of cholesterol to be transported by lymphatics is esterified by a mucosal cholesterol esterase with fatty acids derived from the diet and from mucosal epithelial metabolism. This esterifying system requires bile salts as co-factors (Treadwell & Vahouny, 1968). From the mucosal cell, chylomicron and VLDL cholesterol enter the central lymphatic of the villus. The physical state of cholesterol at this stage is discussed in the section dealing with chylomicra. In many species, the bulk of absorbed cholesterol is carried in lymph in the chylomicron fraction. In rabbits, however, VLDL appears to account for as much as half of absorbed cholesterol (Fraser & Courtice, 1969).

The route of transport of absorbed cholesterol

It is generally considered that cholesterol absorbed from the intestine is exclusively transported by the lymphatic route. Indeed, cholesterol has been used in some absorption studies as a reference substance in order to assess the degree to which a test substance is transported by lymph (see, for example, MacMahon, Neale & Thompson, 1971).

Early work by Mueller (1915) using dogs with thoracic duct fistulae, suggested that lymphatics might account for a major proportion of cholesterol transport, but the possibility that some direct uptake into portal venous blood occurs was not excluded. Bollman & Flock (1951) showed that between 10 and 50% of a fed dose of cholesterol could be recovered in the lymph over twenty-four hours but the data obtained did not indicate whether any portal venous transport occurred. On the other hand, Biggs, Friedman & Byers (1951), using rats with thoracic duct fistulae fed labelled cholesterol, found that at least

90% of the absorbed cholesterol was recovered in the lymph samples and they suggested that, after taking into account the possibility of loss through lymphatico-venous anastomoses, the remainder might also be transported by lymph. Chaikoff *et al.* (1952), using [^{14}C]cholesterol fed to rats with thoracic duct fistulae, concluded that almost all absorbed cholesterol could be recovered in the lymph during a twenty-three hour absorption period, and Hyun *et al.* quoted by Treadwell & Vahouny (1968) considered that all cholesterol transport from the intestine is via lymphatics. Their evidence is obtained from feeding [7-^{3}H]cholesterol to rats with portal venous and thoracic duct fistulae. No detectable counts were found in the portal vein over a period of eight hours and 96–98% of the administered labelled cholesterol was recovered in thoracic duct lymph and in gastro-intestinal tract contents and gut wall.

There is, however, some evidence that other non-polar lipids, including vitamins A and E, may be transported under certain circumstances in small amounts by the portal vein, and it is possible that trivial amounts of cholesterol may also enter the portal vein under special circumstances. MacMahon *et al.* (1971) found that small amounts of radioactivity were excreted in the bile of rats with either mesenteric or thoracic duct fistulae fed [^{14}C]cholesterol. These authors suggest that in animals with bile fistulae re-esterification of free cholesterol in the intestinal mucosa is defective as a result of the lack of bile salts and this allows a greater proportion of free cholesterol to leave the mucosal cell, possibly binding to plasma protein and being taken up to a limited extent by the portal venous blood. Alternatively, the appearance of radioactive isotope in the portal vein might just represent exchange of labelled free cholesterol in plasma with unlabelled free cholesterol in portal blood without any net transport of cholesterol into the portal vein. Such exchange seems a little unlikely and the results would be more in keeping with a small net transport of cholesterol into the portal blood in these animals with bile fistulae.

Quantitative aspects of cholesterol transport
Fat feeding increases small intestinal lymph flow and the entero-hepatic circulation of cholesterol. Since cholesterol is

transported in the lymph, this means that the quantities of sterol passing through the thoracic duct after fat feeding will increase. Clearly, if the fat which is fed contains no cholesterol, the increased lymphatic cholesterol will be of exclusively endogenous origin, i.e. biliary and mucosal.

In studies of a patient with chyluria, Blomstrand & Ahrens (1958) calculated that the total amount of cholesterol reaching the circulation via the thoracic duct each day is between 1 and 3 g. In patients with thoracic duct fistulae, Borgström, Radner & Werner (1970) obtained similar figures ranging between 1.9 and 5.5 g per day. Using the specific activity of labelled cholesterol to measure endogenous and exogenous cholesterol transport, it was found that dietary cholesterol accounts for only a relatively small proportion of the total sterol transported in the thoracic duct lymph in man.

It is clear then, that endogenous cholesterol greatly dilutes exogenous cholesterol in the lumen of the small intestine in man. It is difficult, however, to quantify the proportions of the different forms of endogenous cholesterol (biliary, intestinal mucosal and circulating), which dilute the exogenous cholesterol transported by lymph. In general, it appears that in man the feeding of a high cholesterol diet does not greatly enhance the absolute amounts of cholesterol transported in thoracic duct lymph.

Borgström and his co-workers (1970) have proposed a model for cholesterol absorption and its lymphatic transport which at present would satisfy the data available on this complex subject. In this hypothesis it is proposed that dietary cholesterol mixes in the small intestine with endogenous cholesterol forming a pool, and absorption of a rather constant amount from this combined pool occurs. The proportion of exogenous cholesterol transported in lymph will increase as this becomes a greater proportion of the pool after an increased dietary load of cholesterol. While the dietary part of the pool is small relative to the total pool, there will be no apparent saturation for the absorption of dietary cholesterol. It is possible that the absorptive capacity of the small intestine for cholesterol in man is virtually saturated by the endogenous pool, and that by comparison the dietary contribution to the combined pool is small. This would explain why little change in total cholesterol

Transport of absorbed lipids

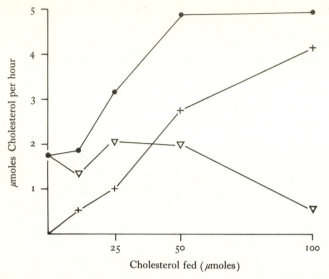

Fig. 7.9. Rate of transport of cholesterol to lymph in thoracic duct-cannulated rats fed 800 μmoles of triolein with different levels of labelled cholesterol. The rates of transport have been calculated as the mean rate of transport per hour between the third and sixth hour after feeding the test meals for total ($\bullet$) and exogenous (+) cholesterol. The endogenous ($\triangledown$) fraction has been calculated as the difference. (Sylven & Borgström, 1968.)

transport occurs in lymph in the thoracic duct when the load of dietary cholesterol is increased. Nevertheless, the *proportion* of exogenous cholesterol in the thoracic duct should rise in these circumstances. In the rat the system is not the same, since by raising the dietary cholesterol load lymphatic cholesterol transport progressively rises to a maximum, and nearly 90% of the cholesterol transported in the lymph is of exogenous origin when high dietary loads are given (Fig. 7.9). These important inter-species differences may be due, at least in part, to differences in the size of the endogenous pool in man and rat.

Phospholipids
In the formation of chylomicra by intestinal mucosal epithelium, phospholipids, in particular lecithin (phosphatidyl choline), are an important stabilising component. Lymph, therefore, could be a route of transport for absorbed phospholipid.

Fig. 7.10. Hydrolysis of lecithin by phospholipase A_2.

Transport of absorbed lipids

Lecithins. Lecithins in the lumen of the small intestine are derived mainly from the diet and bile, though some are also derived from shed mucosal epithelium. In the lumen of the intestine, lecithins are rapidly hydrolysed to yield α-lysolecithin (see Fig. 7.10; Borgström, 1957). Some evidence suggests that rat biliary lecithin may be resistant to this hydrolysis (Boucrot & Clement, 1971). The enzyme responsible for lecithin hydrolysis in the small intestine is pancreatic phospholipase A_2, an enzyme which exists in pancreatic tissue and juice as a zymogen, pro-phospholipase (de Haas, Postema, Nieuwenhuizen & Van Deenen, 1968). This enzyme is activated by tryptic digestion in the small intestine in a manner similar to the activation of the proteolytic series of pancreatic enzymes. The 'cascade' of activation of proteolytic zymogens is initiated by the enzyme enterokinase of the small intestinal mucosa which catalyses the conversion of trypsinogen to trypsin. It is interesting to speculate whether enterokinase deficiency (Hadorn *et al.*, 1971), is associated with defective intraluminal hydrolysis of lecithin as a result of failure of conversion of pro-phospholipase to its active form. In man, further hydrolysis

TABLE 7.3. *Incorporation into the chylomicra of rat thoracic duct lymph of intact dietary 1-Acyl-GPC.* (From Nilsson 1969*b*.)

15 mg 1-acyl-GPC, 25 mg lecithin or 15 mg free fatty acid were fed in 0.5 ml triolein and lymph was collected for 24 hours

Substrate	% recovery of radio-activity in the lymph lipids	Distribution of radioactivity		% of the lecithin label in 1-position
		Lecithin	Glycerides and free fatty acids	
1-[9, 10-^{3}H] pamitoyl-GPC	51–57	26–31	66–73	80–93
[9, 10-^{3}H$_2$] palmitic acid	66–70	1–6	94–98	69–71
[2-oleoyl-9, 10-^{3}H$_2$] lecithin	53–79	5–12	88–95	33–40
[9, 10-^{3}H$_2$]oleic acid	66–72	1–4	96–99	27–46

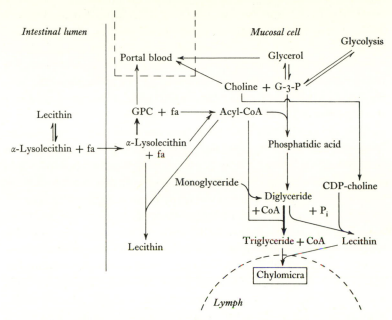

Fig. 7.11. Schematic outline of the metabolism in intestinal epithelium of the products of digestion of lecithin. GPC, glycerol phosphorylcholine; fa, fatty acid; acyl-CoA, acyl-coenzyme A; G-3-P, glycerol-3-phosphate; P_i, inorganic phosphate; CDP, cytidine diphosphate. (From Nilsson, 1969b.)

of α-lysolecithin to glyceryl phosphorylcholine in the gut lumen only occurs to a very limited extent (Arnesjö *et al.*, 1969) but it is hydrolysed to a large extent *in* the mucosa by lysolecithinase activity to yield glyceryl phosphorylcholine and the released fatty acid is subsequently incorporated into chylomicron triglyceride (Nilsson, 1968a). The glyceryl phosphorylcholine thus liberated is not used for lecithin resynthesis. However, a certain amount of absorbed lysolecithin is directly acylated via coenzyme A derivatives of fatty acids in the mucosal cell and this lecithin is thereafter incorporated into chylomicra (Table 7.3; Fig. 7.11).

In the light of this information, earlier studies of phospholipid absorption and lymphatic transport can be readily interpreted. For example, a study by Bloom, Kiyasu, Reinhardt & Chaikoff (1954) showed that a crude phospholipid fraction from the liver of rats fed [^{14}C]palmitic acid, when given to rats with lymphatic fistulae, resulted in the appearance of about 20% of this labelled fatty acid in the lymph as phospholipid. This was interpreted to mean that a significant proportion of phospholipid was being absorbed intact but more probably means that a fairly large proportion of the phospholipid synthesised in the liver had incorporated the labelled fatty acid into the 1-position of the lecithin.

The uptake by the intestinal mucosa of lysolecithin derived from dietary lecithin accounts for only part of chylomicron phospholipid. Scow, Stein & Stein (1967) have shown that at a maximum only 40% of chylomicron lecithin can be derived from α-lysolecithin of dietary origin. What are the other sources of the lecithin of chylomicra? A contribution comes from biliary lecithin, though it is difficult to assess how much this is. Leat & Harrison (1974), on the basis of studies of incorporation of absorbed labelled fatty acids into lymph phospholipids, together with measurements of phospholipid output from bile fistulae in sheep, have suggested that biliary phospholipid accounts for a major part of lymph phospholipid in this species. There is evidence that two other mechanisms can produce lecithin in the mucosal cell which will enter lymph in chylomicra. Stein & Stein (1966) have shown that acylation of lysolecithin derived from blood plasma can occur in the mucosa of the small intestine, and another route involves *de novo* synthesis of lecithin through diglyceride and cytidine disphosphate-choline (Gurr, Brindley & Hübscher, 1965).

As might be expected, during fat absorption mucosal lecithin synthesis increases and this is confirmed by enhancement of the incorporation of ^{32}P into mucosal cell lecithin under these circumstances. Nilsson (1968a) has shown that with increasing amounts of fat fed to rats a greater proportion of absorbed lysolecithin is converted to lecithin, presumably to meet the demand for phospholipid as a stabilising component of chylomicra.

Much of what has been said above implies that chylomicra constitute the most important form of transport of lecithin in lymph. However, considerable amounts of lecithin are present in the clear subnatant fluid obtained after high speed centrifugation of intestinal lymph and this lecithin, bound to other

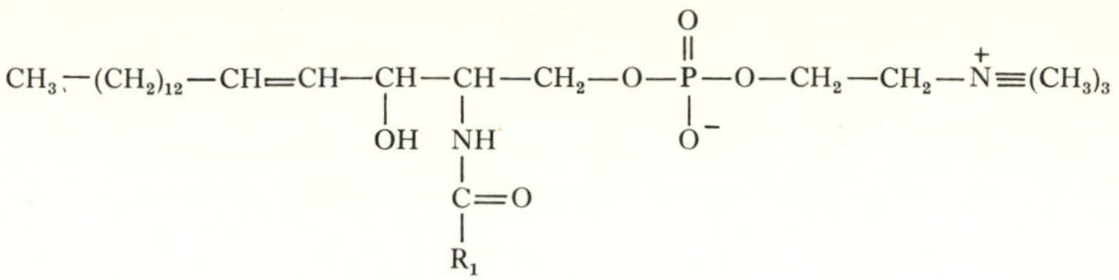

$$CH_3-(CH_2)_{12}-CH=CH-CH-CH-CH_2-O-\overset{\overset{\displaystyle O}{\|}}{\underset{\underset{\displaystyle O^-}{|}}{P}}-O-CH_2-CH_2-\overset{+}{N}\equiv(CH_3)_3$$

(with $\underset{\underset{\displaystyle C=O}{|}}{OH}$ and $\underset{\underset{\underset{\underset{\displaystyle R_1}{|}}{C=O}}{|}}{NH}$ substituents)

Fig. 7.12. Structure of a sphingomyelin.

lipoproteins, probably exchanges, though only to a slight degree, with chylomicron phospholipid. While lecithin is probably transported almost exclusively by intestinal lymph, water-soluble metabolites of lecithin produced by intestinal mucosal epithelium are transported by the portal vein. Saunders, Parmentier & Ways (1968) observed that after feeding choline-labelled lysolecithin to rats, water-soluble radioactive derivatives were transported in the portal vein.

Other phospholipids. There have been relatively few studies of absorption and lymphatic transport of the minor phospholipid species found in the diet. Sphingomyelins (*N*-acyl-sphingosyl-phosphorylcholine) (Fig. 7.12), the only phosphosphingo-lipids found in mammalian tissues, are absorbed by the rat small intestine and are extensively metabolised in the mucosal cells. There is no evidence for the passage of intact sphingo-myelin through the mucosal cells to the lymph. Nilsson (1968*b*, 1969*a*) has shown that enzymes probably located in the micro-

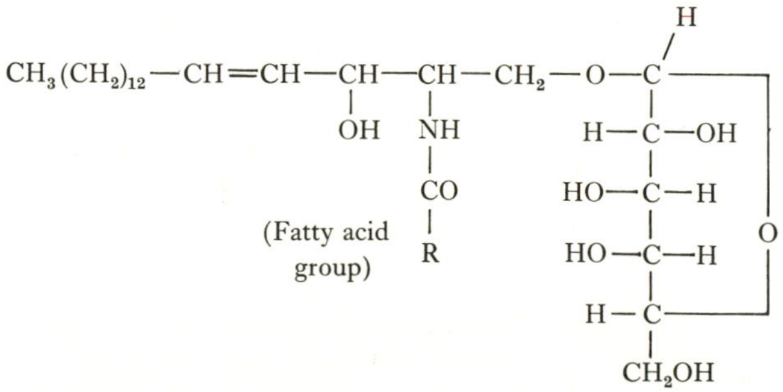

$$CH_3(CH_2)_{12}-CH=CH-CH-CH-CH_2-O-C$$

(Fatty acid group) with OH, NH, CO, R substituents and sugar ring: H-C-OH, HO-C-H, HO-C-H, H-C, CH₂OH

Fig. 7.13. Structure of a cerebroside.

211

Transport of absorbed lipids

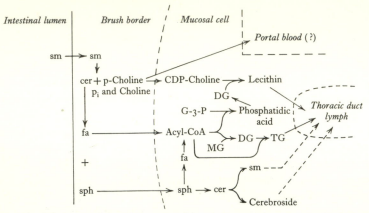

Fig. 7.14. Schematic outline of the metabolism of sphingomyelin in the small intestine. The sphingomyelin is hydrolysed, probably in the brush border, yielding sphingosine, fatty acid and phosphoryl choline or choline and inorganic phosphate. The sphingosine portion is largely converted to fatty acid and enters the chylomicron triglyceride and lecithin but may also take part in biosynthesis of sphingomyelin and cerebroside. A part of the sphingolipids formed in this biosynthesis is incorporated into the chylomicra. Between 10 and 17% of the choline moiety is incorporated into chylomicron lecithin. Abbrevations: cer, ceramide; DG, diglyceride; fa, fatty acid; G-3-P, glycerol-3-phosphate; MG, monoglyceride; TG, triglyceride; sm, sphingomyelin; sph, sphingosine; p-choline, phosphoryl choline; CDP, cytidine diphosphate; P_i, inorganic phosphate. (From Nilsson, 1969b).

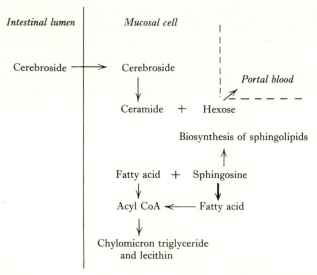

Fig. 7.15. Schematic representation of the metabolism of cerebroside in the small intestine.

212

villous membrane of the mucosal cell release phosphorylcholine and fatty acid portions of the sphingomyelin molecule
from the sphingosine base. The fatty acid released is extensively incorporated into chylomicron triglyceride and lecithin,
and metabolism of the sphingosine base also produces fatty acid
which is similarly handled. Sphingomyelin in lymph, whether
in chylomicra or associated with other lipoproteins, is probably
derived from biosynthesis of the phospholipid in the mucosal
epithelium.

Cerebrosides (glycosphingolipids) (Fig. 7.13) are also absorbed to some extent by the intestinal mucosal epithelium in
the rat and are metabolised during their transit through the
cells. As with sphingomyelin, none of the intact phospholipid
crosses the cell to the lymph. Enzymic digestion releases the
hexose and fatty acid moieties, and the sphingosine fraction
undergoes conversion to fatty acid, as happens in sphingomyelin absorption. These fatty acids are incorporated into
chylomicron triglyceride and lecithin and the carbohydrate is
presumably transported by the portal vein to the liver. Figs.
7.14 and 7.15 are schemes which summarise the processes of
sphingolipid absorption as proposed by Nilsson (1969*b*).

Miscellaneous lipids. In rats, the terpene phytol is partially
absorbed and transported in intestinal lymph partly free and
partly esterified with fatty acids (Baxter, Steinberg, Mize &
Avigan, 1967). Small amounts of hydrocarbons such as hexadecane and octadecane are absorbed into intestinal lymph and
their absorption is enhanced by simultaneous triglyceride
feeding; this is presumably the result of their solubilisation in
mixed bile salt–lipid micelles (Savary & Constantin, 1967;
Borgström, 1974).

Fat-soluble vitamins

Vitamin A

Vitamin A (retinol) (Fig. 7.16) is derived only from animal
sources but plant carotenoids such as β-carotene can be converted *in vivo* to retinol; retinal and retinoic acid are the
aldehyde derivatives of retinol.

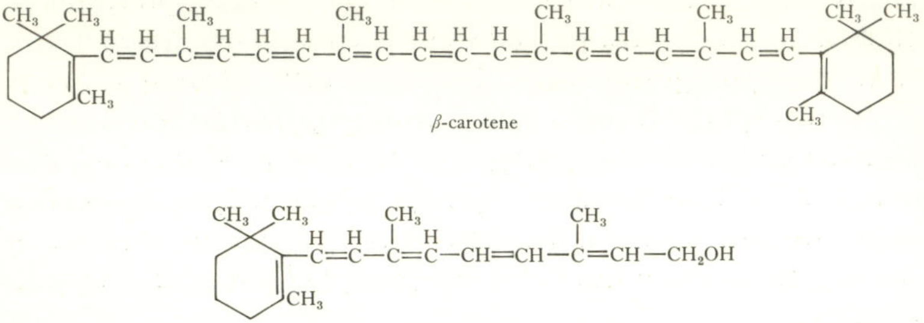

β-carotene

Vitamin A

Fig. 7.16. Structures of vitamin A and β-carotene.

In 1935 Drummond, Bell & Palmer demonstrated lymphatic uptake of vitamin A and carotene in a patient with chylothorax. Forbes (1944) described similar findings with respect to vitamin A absorption in a child with a chylothorax and, in addition, observed that vitamin D absorption also involved the lymphatic route. That the lymphatic route is of major importance in vitamin A transport in man and other species has subsequently been confirmed. For example, Popper & Volk (1944), using fluorescence microscopy, showed the presence of vitamin A in rat intestinal lymphatics after feeding the vitamin. Thompson *et al.* (1950) found that diversion of lymph prevented the appearance of vitamin A in blood or liver after feeding it to rats; and studies in bullocks and sheep (Eden & Sellers, 1949), pigs (Coates, Thompson & Kon, 1950) and guinea pigs (Woytkiv & Esselbaugh, 1951) have yielded similar results.

More recently, investigation of the forms of vitamin A and carotene in lymph has led to the recognition of the intestinal mucosa as a major site for the conversion of β-carotene to vitamin A. Olson (1961) has shown in rats *in vivo* that the upper two-thirds of the small intestine is more active than the lower one-third in the conversion. A cleavage enzyme, carotene 15, 15′ dioxygenase yields retinal which is readily reduced to retinol in the mucosa. Huang & Goodman (1965) examined the lymph of rats with thoracic duct fistulae which were given ^{14}C-labelled retinol or β-carotene. Of the radioactivity recovered in the lymph, 82% was found in the chylomicra after either test substance was given, and 90% of this activity was found as retinyl esters. Only a very small amount of the free alcohol was found in the lymph. The retinyl esters appeared rapidly in the lymph after feeding either β-carotene or retinol, suggesting that this absorptive process was taking place in the proximal small intestine. These authors considered that the lymphatics constitute the only pathway for the transport of the products of vitamin A and β-carotene absorption, since after feeding [^{14}C]carotene virtually no isotope was found in the liver in animals with lymph diversion, although a considerable amount was recovered from the lymph during a forty-eight hour collection. An important observation was that retinyl palmitate was the predominant ester in lymph and that the amounts of saturated retinyl esters (palmitate and stearate, in the ratio 2:1) accounted for more than two-thirds of the labelled esters. Small amounts of retinyl oleate and linoleate were also found. The α position of lymph lecithin is predominantly esterified with palmitic acid (Whyte, Goodman & Karmen, 1965). It may be that in the mucosal cell, a transesterification process occurs between vitamin A and the α position of lecithin.

Similar studies have been performed in patients with thoracic duct fistulae (Goodman *et al.*, 1966). Labelled vitamin A, vitamin A acetate and β-carotene were given by mouth. The radioactivity appeared mainly within three to ten hours and 70% of this was found in washed chylomicra. Fig. 7.17 illustrates the time course of absorption of β-carotene and its products into thoracic duct lymph in two human subjects. After giving vitamin A or β-carotene, most of the activity appeared as retinyl esters whose fatty acid composition was rather constant, being similar to that of the α position of lymph lecithin, and independent of dietary fatty acid composition. These results are comparable with those obtained for the rat. Furthermore, the ratio of saturated to unsaturated fatty acids was closely similar to that found in the rat. Small amounts of β-carotene appeared unchanged in the lymph; this contrasts with the absence of β-carotene in rat lymph during its absorption. Vitamin A acetate seemed to be entirely hydrolysed and re-esterified with long-chain fatty acids during absorption. Another study in man yielded very similar results (Blomstrand & Werner, 1967).

Fidge, Shiratori, Ganguly & Goodman (1968) have shown

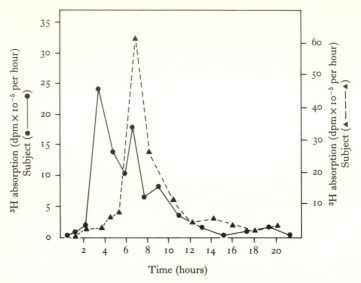

Fig. 7.17. The time course of the absorption of radioactivity into thoracic duct lymph in two human subjects fed β-carotene $-15, 15'-{}^3$H. (Goodman *et al.*, 1966.)

that retinal given to rats is reduced to retinol which is subsequently esterified prior to lymphatic transport. On the other
hand when labelled retinoic acid was given, little of the label
appeared in lymph. The free acid and more polar metabolites
were found in bile and it would seem likely that retinoic acid
is transported by the portal venous route from the small
intestine.

Despite the wealth of evidence regarding the function of
lymph vessels in vitamin A transport, there are some sugges-
tions that vitamin A may be carried by portal venous blood
under certain circumstances. Murray & Grice (1961) found that
ligation of the thoracic duct in rats did not affect the appearance
of vitamin A in the liver; lymphatico-portal anastomoses did
not appear to explain these results. That the portal blood could
transport absorbed vitamin A in the presence of thoracic duct
obstruction was further suggested by elevated portal:aortic
blood ratios of the vitamin. Retinyl esters probably have to be
hydrolysed before absorption (Mahadevan, Seshadri Shastry
& Ganguly, 1963). Retinyl palmitate fed to rats appears to be
extensively hydrolysed and re-esterified during its passage

across the intestinal mucosa (Lawrence, Crain, Lotspeich & Krause, 1966). Considerable amounts of the vitamin appeared in livers of rats with thoracic duct fistulae, after feeding vitamin A palmitate, suggesting that other routes might play a part in the transport of some of the vitamin. In the absence of bile, vitamin A is poorly absorbed by rats but lymph:liver ratios in these animals indicate that the portal venous blood might carry some vitamin A (Gagnon & Dawson, 1968). It seems that there exists a partition of vitamin A between lymph and portal blood, normally favouring lymph. Under certain circumstances, the portal venous route may play an important part. As in rats, absence of bile in the intestine in man greatly reduces vitamin A absorption and perhaps diverts transport from the lymphatic to the portal venous route (Forsgren, 1969). Bile salt micelles, in addition to acting as a vehicle for vitamin A transport to the mucosal epithelial cell, also present esters of vitamin A in a form which is readily attacked by a pancreatic esterase (Erlanson and Borgström, 1968). Bile salts appear to act as co-factor for this enzyme.

Vitamin D

The two most important biological forms of this vitamin are the sterols, vitamin D_2 and vitamin D_3 (Fig. 7.18) which are derived from ergosterol and 7-dehydrocholesterol by the action of ultra-violet irradiation. Vitamin D_3 occurs naturally in high concentrations in fish liver oils, and vitamins D_2 and D_3 are added to fortify certain foods.

In 1963, Schachter, Finkelstein & Kowarski described the biosynthetic preparation of ^{14}C-labelled vitamin D_2; they demonstrated that within an hour of an oral dose of this preparation to rats the lymph contained the vitamin, mainly as the free sterol incorporated into chylomicra. In this same study, they demonstrated that when vitamin D_2 was injected intravenously it was taken up by the small intestinal mucosa and liver; it was metabolised to some extent in the liver and significant amounts of water-soluble derivatives appeared in bile. These same authors in 1964 described similar studies which included the use of ^{3}H-labelled vitamin D_3. The jejunum appeared to be the principal site of absorption of the vitamin

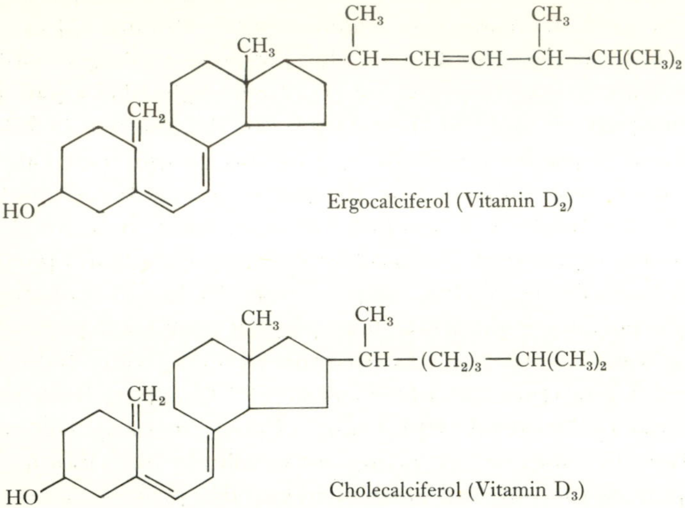

Fig. 7.18. Structures of vitamins D_2 and D_3.

and in intestinal lymph it was found mainly as free biologically active sterol in the chylomicra. Between 80 and 97% of the absorbed radioactivity was recovered in lymph. Vitamin D_3 appeared to be absorbed more rapidly into lymph than vitamin D_2. They also showed that bile salt, and in particular sodium taurocholate, was required for optimal absorption of vitamin D. While uptake of the sterols by the intestinal mucosa appeared to be quite rapid, the transit into lymph was a slower process, possibly dependent on mucosal lipoprotein synthesis. Bell (1966), using rats with thoracic duct fistulae to compare the esterification which occurred during absorption of labelled vitamin D_3 and cholesterol, found that most of the absorbed cholesterol was esterified with saturated and mono-unsaturated fatty acid while very little vitamin D_3 was esterified. Fraser & Kodicek (1968), using vitamin D_3, found that 43% of an oral dose could be recovered from the rat's thoracic duct in twelve hours. Of this, only 1.4% was esterified, the main fatty acids being palmitate (31%), stearate (25%), oleate (16%) and linoleate (16%).

The handling by the gastro-intestinal tract of vitamin D esters, a natural form of the vitamin in certain fish liver oils, has been studied in rats by Bell & Bryan (1969). After administration of labelled vitamin D_3 and vitamin D_3 oleate, they found vitamin D ester, free vitamin and other more polar derivatives in the thoracic duct lymph; the free sterol predominated. Lesser amounts of the free vitamin D were recovered in the lymph after feeding the ester than after giving the free sterol. In-vitro studies showed that pancreatic enzyme preparations had a limited ability to hydrolyse vitamin D esters. These experiments do not determine whether hydrolysis of vitamin D esters in the lumen of the small intestine is an obligatory step in the absorption of this form of vitamin D, as is the case for cholesterol esters. If hydrolysis is necessary, the poorer absorption after giving the ester might reflect the limited affinity of pancreatic enzymes for these esters.

In a study of the influence of mixed bile salt–lipid micelles on the absorption by rats with cannulated intestinal lymphatics of cholesterol and vitamin D_3, Thompson, Ockner & Isselbacher (1969), found that the appearance of the sterols in intestinal lymph was increased by the inclusion of either palmitic or linoleic acid in the micelles. In rats with intact lymphatics, on the other hand, it was found that the uptake of sterols from the intestinal lumen was the same whether the bile salt micelles included fatty acid or not. This suggested that the effect of micellar lipid in enhancing cholesterol and vitamin D appearance in intestinal lymph is through an action on the transport of these sterols *out of* the mucosal cells into lymphatics.

The transport form of absorbed vitamin D in man appears to be similar to that found in rats (Blomstrand & Forsgren, 1967). Data from patients with thoracic duct fistulae showed that about 50% of an oral dose of labelled vitamin D_3 could be recovered from lymph lipids in eighteen hours, most of this was associated with the chylomicra as the free sterol; bile appeared to play an essential part in its absorption and in a series of investigations of the intestinal absorption of labelled vitamins A, D, E and K, Forsgren (1969) showed that in the presence of biliary obstruction in man, the uptake of these vitamins into thoracic duct lymph is severely impaired (see, for example, Fig. 7.19).

As with other fat-soluble vitamins, it is possible that under

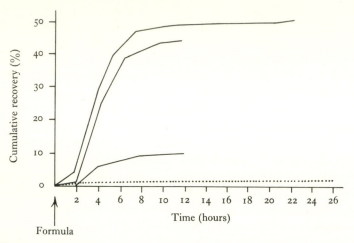

Fig. 7.19. Cumulative recovery of radioactivity in thoracic duct lymph in patients with biliary obstruction (·····) and controls (——) after feeding 0.2 mg labelled vitamin $D_3-1,2-{}^3H_2$. (Forsgren, 1969.)

certain circumstances, vitamin D may be transported by the portal venous route. In man, vitamin D_3 absorption from bile salt micellar solution infused into the duodenum is greater when the micelles include octanoic acid than when they include oleic acid; it is possible that the medium chain fatty acid may be responsible for diverting some vitamin D into the portal vein (Pihl, Iber & Linscheer, 1970). In the liver, absorbed vitamin D_3 is hydroxylated to 25-hydroxy vitamin D_3 and this appears in the circulation and also undergoes enterohepatic circulation. The route of transport of this compound from the intestine is not known.

Vitamin E

The tocopherols, the vitamins E, of the diet occur mainly in plant materials, the richest sources being seed oils and wheat germ oil. Fig. 7.20 shows the structure of α-tocopherol, the most biologically active of the tocopherols. Dietary vitamin E mixes with endogenous vitamin E in the small intestine. Studies by Klatskin & Molander (1952) have shown that human bile contains vitamin E in concentrations comparable with those in plasma. This vitamin, therefore, undergoes entero-

α-tocopherol

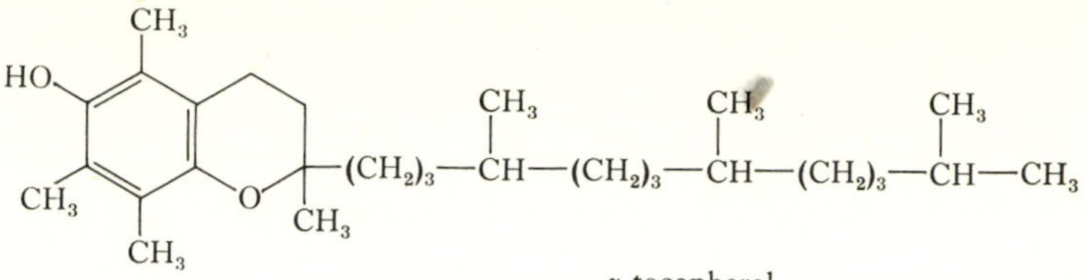

Fig. 7.20. Structure of α-tocopherol.

hepatic circulation. As with other lipid species, vitamin E and its derivatives, prior to absorption, are solubilised in the intestinal lumen in undigested neutral lipids – the oil phase – and in bile salt–lipid micelles.

In the rat, α- and γ-tocopherol are absorbed from the gut unchanged; intestinal lymph acts as the major transport route (Johnson & Pover, 1962; Peake, Windmueller & Bieri, 1972). In man also, α-tocopherol appears in lymph in the unesterified form (Blomstrand & Forsgren, 1968a). As with other fat-soluble vitamins there are indications that some tocopherol may be transported from the intestine by the portal vein. MacMahon *et al.* (1971) have shown that the bulk of absorbed α-tocopherol is transported in the lymph and when the labelled vitamin is given by intraduodenal infusion in bile salt micellar solution, together with exogenous lipids, it is chiefly transported as a component of chylomicra. When the vitamin is incorporated into pure bile salt micelles, a much smaller proportion is transported in chylomicra, the remainder probably being carried on very low density lipoproteins. These workers also found that up to 8 % of the label could be recovered from bile and urine, despite diversion of mesenteric or thoracic duct lymph. Such results suggest that some of the vitamin might be transported by the portal vein and its tributaries, and this idea is strengthened by the observation that concentrations of [^{3}H]α-tocopherol in portal venous plasma were higher than in aortic plasma during absorption.

Bile and pancreatic secretions play an important part in the absorption of tocopherols. Gallo-Torres (1970a, b) administered α-tocopheryl acetate and nicotinate intragastrically to rats in an emulsion containing monoglyceride, carbohydrate

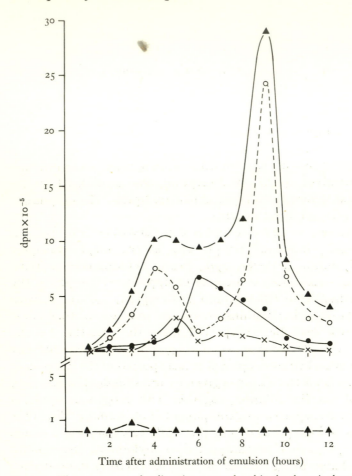

Fig. 7.21. The appearance of radioactive α-tocopherol in the thoracic duct lymph of rats at hourly intervals in the presence of biliary and pancreatic secretions (upper graph) or in their absence (lower graph). In both instances, an emulsion was administered containing protein, carbohydrate, mono-olein and [³H]α-tocopheryl acetate. Total α-tocopherol (all forms), ▲——▲; unesterified α-tocopherol, O----O; α-tocopherol-p-quinone, ●——●; α-tocopherol acetate, ×——×. Each point on the graph represents the mean of three experiments. (Gallo-Torres, 1970a.)

and protein. Most of the vitamin appeared in the lymph in the unesterified form indicating a hydrolytic step during absorption. In the absence of bile or pancreatic juice, very little absorption occurred (Fig. 7.21). In this experiment, pancreatic secretion was probably responsible for the hydrolysis of the vitamin E esters. Where substantial amounts of triglyceride are present during absorption of the vitamin, a second role for pancreatic juice could be the formation by pancreatic lipase of the polar products of lipolysis of triglyceride, i.e. mono-glycerides and fatty acids, which – by amphiphilic solubilisa-tion with bile salts as mixed micelles – increase the solubilising capacity of the bile salts for the non-polar vitamin E and its esters, and thereby facilitate their absorption. Other unknown factors must be important, however, since Muralidhara & Hollander (1977) have reported that the inclusion of poly-unsaturated fatty acids in bile salt micelles appears to depress α-tocopheral uptake from these micelles by rat small intestine. It is interesting that α-tocopheryl nicotinate given to chickens is absorbed primarily as the ester and appears as such in plasma

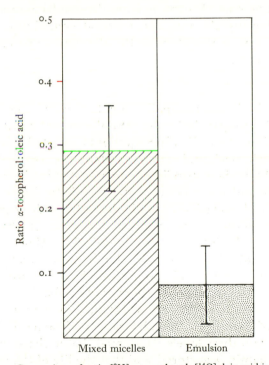

Fig. 7.22. Comparison of ratio [³H]α-tocopherol: [¹⁴C]oleic acid in lymph after intra-duodenal infusion of a mixed micellar solution or a 2.5 mM taurocholate emulsion to lymph–bile fistula rats. Standard deviations are shown by the horizontal lines. (McMahon & Thompson, 1970.)

and liver Gallo-Torres (1970*b*). In these birds, it appears that hydrolysis does not precede absorption and it may be that the ester, once absorbed, is transported in chickens by the portal vein.

Several studies indicate the importance of bile salts in vitamin E absorption. In human subjects, Forsgren (1969) found that biliary obstruction greatly reduces the absorption of vitamins A, D, E and K as judged by their transport in thoracic duct lymph after oral administration, and MacMahon & Thompson (1970) have shown in rats with bile diversion that while a polar lipid, such as oleic acid, is absorbed into mesenteric lymph almost as well from an emulsion as from bile salt micelles, the non-polar α-tocopherol was poorly absorbed from the emulsion when compared with the micellar solution (Fig. 7.22). The poor absorption from the emulsion was associated with low uptake into the mucosal epithelium which suggests that it is at this stage in the absorptive process that bile salts are acting.

Vitamin K

Dietary vitamin K is principally found in green vegetables. The vitamin K group includes the naturally-occurring vitamins K_1 (phylloquinone) and K_2 (farnoquinone), isolated originally from putrid fish meal, and synthetic vitamin K_3 (menadione) (Fig. 7.23). While the naturally-occurring vitamins K_1 and K_2 are fat-soluble and insoluble in water, synthetic forms, such as menadione, which lack the long hydrocarbon side chain, are slightly water-soluble. The route of transport of absorbed vitamin K from the intestine was described by Mann, Mann & Bollman (1949) who studied development of blood coagulation defects in rats due to hypoprothrombinaemia soon after the establishment of a chronic thoracic duct fistula. Normal plasma prothrombin levels were maintained by parenteral administration of vitamin K, though large transfusions of plasma failed to maintain these levels. It seemed that in these animals little of the vitamin was stored and that the life of the plasma prothrombin was short. The authors concluded that the lymphatic system was the principal route of absorption of this vitamin. Jaques, Miller & Spinks (1954), using ^{14}C-labelled

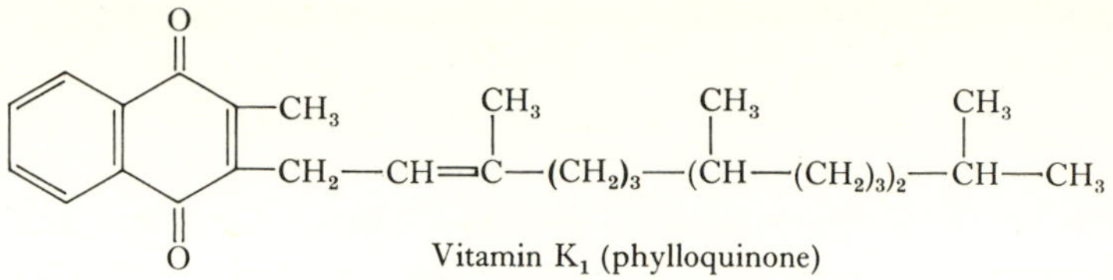

Vitamin K_1 (phylloquinone)
2-methyl-3 phytyl-1,4-naphthoquinone

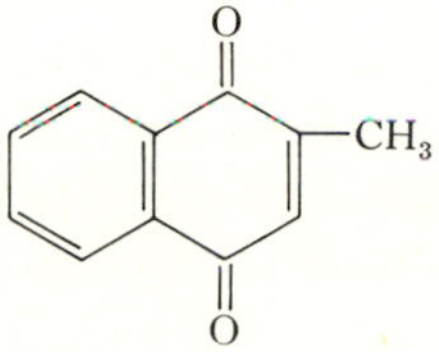

Vitamin K_3 (menaquinone, menadione)
2-methyl-1,4-naphthoquinone

Fig. 7.23. Structures of vitamins K_1 and K_3.

vitamin K preparations, found that approximately 50% of a dose of vitamin K_1 and all of a dose of vitamin K_3 was absorbed by rats. Vitamin K_1 appeared to be transported mainly by the intestinal lymphatics, while vitamin K_3 seemed to be transported by the portal venous route. With biliary diversion, little vitamin K_1 was absorbed, but absorption was restored by the addition via the duodenum of normal amounts of bile, together with a pancreatic lipase preparation. This last observation may indicate that the presence of the products of lipolysis, fatty acid and monoglyceride, play an important role in solubilising vitamin K_1 prior to its absorption. Bile diversion did not seem to affect absorption of vitamin K_3. Both vitamins appeared to be extensively recycled through bile.

These findings regarding the route of transport of vitamin K_3 were re-examined by Mezick, Tompkins & Cornwell (1968) who measured ratios of radioactivity in portal venous and aortic blood after administration of [^{14}C]menadione to rats. Their results suggested that a significant proportion of this vitamin preparation may be transported by the lymphatic route. Similar portal:aortic ratios were found in dogs, and thoracic duct lymph was found to transport 9.6–12.3% of the absorbed vitamin. They also demonstrated that the hepatic lymph does not make a large contribution to this figure, although during the early phase of absorption, some of the

225

vitamin may be delivered to the thoracic duct by this route.

In man, Blomstrand and Forsgren (1968*b*) have made a careful study of the absorption of ^{3}H-labelled vitamin K_1 in patients with thoracic duct fistulae. They recovered, within twenty-four hours, 19.1–61.5% of the vitamin, most of this appearing within eight hours of administration. The radio-activity was found as unaltered vitamin, mainly in the chylomicra. In the presence of obstructive jaundice, the absorption was reduced to less than 3% of the administered dose. This observation is similar to that made by Jaques, Millar & Spinks (1954) in the rat, regarding the effects of biliary diversion on vitamin K absorption.

These experiments throw a great deal of light on the mechanisms of absorption and transport of dietary vitamin K in man and rats. It is possible that absorbed vitamin K excreted in the bile may be handled in a similar fashion, though metabolites of dietary vitamin K which are more polar than their parent compounds may, after being excreted in the bile, be absorbed to a significant extent into the portal venous system.

Vitamin K_2 is a metabolic product in many bacteria including some species present in the intestine of several higher animals. Such bacterial production is responsible for the high quantities of this vitamin in putrefying fish meal. Coprophagic animals, such as rats, obtain much of their vitamin K from their faeces and it is likely that this vitamin, synthesised in the colon is not extensively absorbed in the colon but is only absorbed after ingestion of the faeces. Man, however, appears to derive the vitamin from intestinal bacterial activity; deficiency in man can only be induced by gross dietary restriction and antibiotic alteration of the gut flora (Frick, Riedler & Brögli, 1967). Most of the bacterial production of vitamin K_2 is in the large intestine. It is possible that some absorption of the vitamin by the large intestine can occur in man. However, the fact that the terminal ileum has a bacterial population makes it more probable that bacterial vitamin K is absorbed in this region where bile salts may assist in its solubilisation.

Ubiquinones

The ubiquinones, a group of compounds related to the vitamin E and K group and which include coenzyme Q, an electron transport compound found in mitochondria, behave in a similar fashion to other fat-soluble vitamins. Using phytylubiquinone (hexahydroubiquinone-4) as a model compound for this group, Blomstrand & Gürtler (1971) have demonstrated in man that this compound enters lymph intact, becoming incorporated into chylomicra. As much as 80% of the dose of phytylubiquinone given in a test meal was recovered in thoracic duct lymph in twenty-four hours. Although the percentage absorption of a dose of coenzyme Q_{10} in rats is relatively small it appears that this compound is chiefly transported by thoracic duct lymph (Katayama & Fujita, 1972*a*). 80% of this is incorporated into chylomicra and 10% into VLDL. It is interesting that no correlation could be found between lymph flow and cumulative amount of the compound transported over forty-eight hours. This suggests that the capacity of the intestine to absorb the compound limits its appearance in lymph. Where the transfer of a compound across the mucosal epithelium is fast, however, augmentation of lymph flow may increase its lymphatic transport.

Summary

Dietary lipids are extensively digested in the small intestine. The hydrolysis is chiefly accomplished by pancreatic hydrolases which cleave fatty acid ester bonds. Bile released during fat digestion, contributes to the process by providing bile salts which stabilise emulsions of undigested and partially digested glycerides and solubilise fatty acids and partial glycerides, together with non-polar lipids such as sterols and fat-soluble vitamins. The solubilising unit is the bile salt micelle. It is generally believed that these micelles are of greatest importance to the absorption of highly non-polar lipids. The solubility of these substances in bile salt micelles is greatly enhanced by the inclusion of fatty acids and monoglycerides, and this may partly explain the observation that normal pancreatic digestive function is required for optimal rates of absorption of fat-soluble vitamins. Bile itself contains lipids which undergo an entero-hepatic circulation, and where a particular lipid, such as cholesterol, is present in relatively small amounts in the diet, the biliary (endogenous) contribution to the pool of the substance in the lumen of the small intestinal lumen, becomes quantitatively important.

The absorption of the various lipids is accomplished by a non-energy-dependent process though metabolic changes in these lipids within the cell facilitate their absorption. Re-synthesis of triglyceride, as well as esterification of sterols, lysolecithin, and some fat-soluble vitamins occurs within the small intestinal epithelium. Synthesis of lipoproteins and phospholipids within the cell is necessary for the formation of chylomicra which are the principal vehicle for transport of absorbed lipid to the systemic circulation. These chylomicra leave the cell through its lateral cell membrane by an ill-defined mechanism, traverse the lamina propria of the intestinal villi to enter the central lymph vessel. It appears that the portal venous route plays a less important role in fat absorption, being chiefly concerned with transport of unesterified short- and medium-chain fatty acids bound to protein. Only minor amounts of other lipid species such as long-chain fatty acids are transported in portal venous blood, but there is evidence that these amounts increase when re-esterification of lipids in the mucosal cell is defective.

8 LIVER LYMPH

The anatomy of lymph formation in the liver

Any consideration of hepatic lymph formation must take into account the special microvascular anatomy of the liver since the structure of this system accounts for many of the observations made on liver lymph flow and composition. The anatomical arrangement of the low pressure hepatic sinusoid is diagrammatically illustrated in Fig. 8.1 which shows the important space of Disse, considered to be the main extracellular compartment from which liver lymph is derived. This compartment is not recognisable in ordinary histological preparations of the liver but can be readily seen in certain pathological conditions such as hepatic venous obstruction (Bolton & Barnard, 1931). The lining of the hepatic sinusoid consists of endothelium and the phagocytic Kupffer cells, the 'littoral' cells. Plate 8.1 shows the ultrastructure of the endothelial cells. The

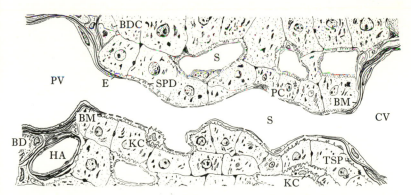

Fig. 8.1. Diagrammatic representation of a rat liver sinusoid. Abbreviations: S, sinusoid; PV, portal vein; CV, central vein; BM, basement membrane; E, endothelium; SPD, perisinusoidal space of Disse; KC, Kupffer cell; PC, parenchymal cell; BD, bile duct; BDC, bile duct cell; HA, hepatic artery; TSP, tissue space. (Redrawn from Burkel & Low, 1966.)

229

porous nature of the sinusoidal endothelium has been extensively studied by electron microscopy. Discontinuities in the lining of various descriptions including 'gaps', 'slits', 'pores' and fenestrae have been found. Their diameters are generally estimated to be 100–200 nm. For a detailed discussion the reader is referred to the paper of Wisse (1970). In the space of Disse occasional strands of basement membrane are found associated with collagen fibrils. Through the highly permeable sinusoidal wall there is ready access of the chemical constituents of blood plasma to the surface of the parenchymal cell via the space of Disse. Liver lymph derived from the interstitial fluid in the space of Disse is collected principally by the periportal lymph vessels, though a minor proportion is drained by lymph vessels related to the centrilobular vein. (see Chapter 2).

Lymphatic vessels as such are not identified in the region of the sinusoids but the initial lymph vessels of the liver are found in portal tracts and in relation to the central vein of the hepatic lobules. It appears that, whereas sinusoidal blood is flowing towards the centre of the hepatic lobule, interstitial fluid in the space of Disse is flowing in the opposite direction to the peripherally situated portal tracts. While the major part of the liver is perfused with blood in hepatic sinusoids, a peribiliary capillary plexus exists in the portal tracts, and this circulation is comparable with the microcirculation of other tissues. It is supported by the connective tissue of the portal tracts and supplied by hepatic arterial blood so pressure in these vessels is higher than in the sinusoidal vessels (Brauer, 1963). The route of drainage of blood in these portal tract capillaries is not clear though it is likely that their blood enters the sinusoids. Lymph derived from fluid in the space of Disse is probably modified by addition of lymph derived from this portal micro-circulation. Approximately 80% of hepatic lymph leaves the liver by the hilar lymphatic system to enter the cisterna chyli via the main hepatic lymph vessel(s) (Ritchie, Grindlay & Bollman, 1959). The remaining 20%, in lymph vessels associated with the central veins, does not go to the cisterna chyli–thoracic duct system but passes in association with the hepatic veins into retrosternal lymphatics which run to the root of the neck along the course of the internal mammary artery, to enter great veins there.

It is thus clear that liver lymph largely originates from a rather specialised type of interstitial fluid which is probably in free exchange with plasma as far as its non-particulate constituents are concerned. There is no information about the relationship between the space of Disse and other interstitial spaces of the liver, if indeed such spaces exist. Burkel and Low (1966) consider that the space of Disse is in free communication with typical tissue spaces surrounding the central vein of the liver lobule and in the portal tract (see Fig. 8.1).

The flow of liver lymph

It is difficult to arrive at more than a very rough approximation of resting lymph flow rates from any organ, partly because of wide variation in normal flows and also because it is seldom possible to ensure a complete collection of all lymph draining an organ. This is true for the liver since direct measurement by complete lymph collection would require cannulation of hilar lymphatics and the lymph vessels associated with the hepatic vein. In spite of the difficulty in cannulating hepatic lymph vessels, there is a considerable amount of information on this subject.

In general, the liver and intestine contribute the major part of thoracic duct lymph and 25–50% of thoracic duct lymph is estimated to come from the liver in the dog (Cain *et al.*, 1947). Nix *et al.* (1951) concluded that 50% of thoracic duct flow at rest in dogs is derived from the liver. In the cat, Morris (1956*a*) has found that liver lymph contributes about 30% of total thoracic duct lymph (Table 8.1). Hepatic lymph flow in the rat is difficult to study, but the data of Bollman, Cain & Grindlay (1948) – 5 ml per twenty-four hours – and Friedman, Byers & Omoto (1956) – 0.45–3.5 ml per twelve hours agree quite well. Mann & Higgins (1950) found hepatic lymph flow in the rat to average 2.0 ml per day. In calves, Shannon & Lascelles (1968*a*) measured flow rates of 0.64 ml per kg body weight per hour from the hepatic duct and in anaesthetised rabbits, del Rio Lozano & Andrews (1966) observed hepatic

TABLE 8.1. *Flow rate and protein concentration of thoracic duct, liver and intestinal lymph of the anaesthetised cat (nembutal).* (From Morris, 1956a.)

Lymph	Number of animals	Flow (ml per kg per hour)	Protein (g per 100 ml)
Thoracic duct	66	2.42±0.12	4.53±0.1
Hepatic duct	38	0.73±0.06	6.06±0.05
Intestinal duct	30	1.54±0.1	4.19±0.1

All values are means±standard error of mean.

lymph flows of 0.8–12 ml per hour. As a rough approximation, lymph flows from the liver are generally in the range of 0.4–0.6 ml per kg liver per min (Brauer, 1963). Total hepatic blood flow is approximately 1 litre per kg liver per min.

The formation and composition of liver lymph

The highly permeable nature of the liver sinusoidal wall was discussed by Starling (1909) and confirmed by the finding that large molecular weight dextrans and small spherical particles could be recovered from hepatic lymph after intravenous injection (Grotte, Juhlin & Sandberg, 1960; Mayerson, Wolfram, Shirley & Wasserman, 1960). Plasma proteins readily pass into the space of Disse (Courtice, Woolley & Garlick, 1962), and this accounts for the high concentration of protein, approaching that of plasma, in hepatic lymph. Since all plasma proteins pass rather freely across the sinusoidal wall, the oncotic pressure effect of intravascular protein, which in other tissues depends on the relative impermeability of capillary walls to plasma proteins, is largely lost in the microcirculation of the liver. The balance of fluid movements across this wall will clearly depend almost entirely on hydrostatic pressure gradients. A small rise in pressure in the hepatic vein of 0.75–1.5 mmHg was found to cause a transudation of fluid from the capsular surface of the isolated perfused rat liver (Brauer, Holloway & Leong, 1959). In other capillary beds, much greater rises in venous pressure are required before extra interstitial fluid is formed. In view of this, the raised venous pressure of congestive cardiac failure can be expected to enhance interstitial fluid formation in the liver and hepatic lymph flow, and the increase in the hepatic interstitial fluid compartment is likely to be partly responsible for hepatomegaly in this condition.

The protein content of liver lymph

As already mentioned, the most striking feature of hepatic lymph is its high protein content. Estimates of lymph:serum ratios for various proteins range down from 0.9 for smaller proteins to 0.5 for very large proteins (Table 8.2; Dive, Nadalini & Heremans, 1971).

The source of hepatic lymph proteins has received a great deal of attention since it has been proposed by some workers that liver lymphatics may be the route of delivery of a proportion of newly synthesised protein from the liver to the bloodstream (CoTui, Barcham & Shafiroff, 1944). This would be of particular importance in studies of the kinetics of plasma protein metabolism. In 1962, Kolmen & Vita concluded that some fibrinogen reaches the plasma by primary hepatic lymphatic transport. This was based on studies of the recovery of plasma fibrinogen levels after bleeding in dogs with lymph fistulae. The more rapid recovery of fibrinogen levels when thoracic duct lymph was re-directed to the blood was adduced

TABLE 8.2. *Relationship between the concentrations of the different proteins in hepatic lymph and serum.* (From Dive *et al.*, 1971.)

Protein	Molecular weight	Number of dogs	Lymph : serum ratio
Orosomucoid	44 000	9	0.874±0.061
Albumin	69 000	12	0.833±0.035
Transferrin	90 000	12	0.773±0.055
IgG$_{2ab}$	160 000	12	0.650±0.074
α_2Macroglobulin	820 000	12	0.506±0.056
IgM	1 000 000	12	0.471±0.049

Lymph:serum ratios are means±standard error of mean.

as evidence for a lymphatic route of delivery of the newly synthesised protein. This is only a rather indirect means of studying this problem. Furthermore, lymph drainage will inevitably deplete the animal of large amounts of all plasma proteins since there is a very considerable transport of plasma proteins through the lymphatics returning these macromolecules to the blood circulation. A more satisfactory approach to this problem is the study of concentrations of tagged newly synthesised proteins in hepatic lymph and blood plasma. Such experiments have been carried out by Alper, Peters, Birtch & Gardner (1965) using [^{14}C]glycine and following the fate of newly biosynthesised haptoglobin, fibrinogen and γ-globulin in dogs. Measurements of specific activity of these proteins in hepatic venous plasma, portal venous plasma and thoracic duct lymph show that haptoglobin and fibrinogen reach the blood circulation directly; the rapid appearance of newly labelled γ-globulin in lymph at a specific activity comparable with plasma is in keeping with the idea that this protein, which is derived from extrahepatic sites, may, at least partly, enter the blood circulation through non-hepatic lymphatics including those draining the intestinal mucosa (see chapter 6).

As already mentioned most studies indicate that hepatic lymph protein is derived from the plasma by filtration. For example, in the rabbit, Woolley & Courtice (1962) using labelled albumin and γ-globulin concluded, on the basis of serum and hepatic lymph specific activities, that hepatic lymph proteins are largely filtered from plasma in the porous hepatic microvasculature. Smallwood *et al.* (1968) have come to a similar conclusion in a study of the appearance of newly synthesised [^{14}C]fibrinogen and [^{14}C]albumin in the thoracic duct in dogs. The proportion of biosynthesised [^{14}C]albumin and intravenously injected ^{131}I-labelled albumin recovered in thoracic duct lymph were closely comparable and these authors proposed that newly synthesised albumin in the liver enters the space of Disse and from this pool will be drained by venous blood or lymph at a rate proportional to their rates of flow. Fig. 8.2 is a diagrammatic representation of this hypothesis which assumes free movement of albumin molecules across the sinusoidal endothelium. It is interesting that in this study partial

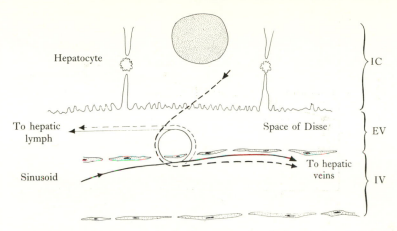

Fig. 8.2. Diagram showing the proposed mixing pool in the hepatic sinusoidal Space of Disse. IC, intracellular; EV, extravascular; IV, intravascular. Newly synthesised albumin is shown by the broken line and circulating albumin by the solid line. Heavy lines indicate major transport routes. (From Smallwood *et al.*, 1968.)

obstruction of hepatic venous outflow increased the amount of labelled albumin transported by the hepatic lymph though the increase could be explained by increased lymph flow, which was approximately doubled by this manoeuvre. Dive *et al.*, (1971) also using dogs have compared the concentrations of a number of plasma proteins in hepatic lymph and serum (Table 8.2). It is apparent that there is some selectivity in the permeability of the hepatic microvasculature to large proteins such that very large proteins are present in relatively low concentration in hepatic lymph compared with serum. Fig. 8.3 shows the relationship between the molecular weight of a protein and the ratio of its lymph : serum concentration to the lymph : serum concentration ratio for albumin. It can be seen that at very high molecular weight there exists a bulk transfer which is independent of the size of the protein species being transferred, indicated by the horizontal broken line, while a molecular sieving also exists (Dive *et al.* 1971). This sieving, based on molecular size, involves a pore system for which these authors have computed a mean radius of 102 A (10.2 nm). The anatomical basis for the restraint on passage of large protein molecules into hepatic lymph is not clear since there are apparently

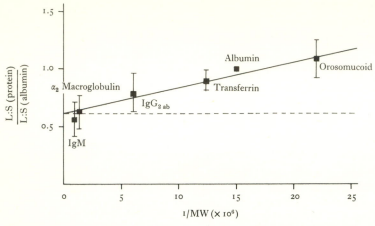

Fig. 8.3. Relationship between the ratio of lymph:serum (protein) to lymph:serum (albumin) and molecular weights of the proteins. There is a linear relationship between this ratio and the inverse of the molecular weight but the intersection of the ordinate (------) suggests two forms of transfer. The constant term of the equation:

$$\frac{\text{Lymph:serum (protein)}}{\text{Lymph:serum (albumin)}} = 0.606 + \frac{0.0216 \times 10^6}{\text{MW}}$$

corresponds to the process of bulk transfer without modification of the protein composition. (Dive *et al.*, 1971.)

many fenestrae in the sinusoidal endothelial cells and gross deficiencies in their basement membrane. There are a number of possible explanations. First, it might be that electron microscopic appearances do not accurately reflect the true anatomical situation, and that there is a molecular sieving at the sinusoidal wall. A more likely explanation is that one cannot equate the fluid in the space of Disse with hepatic lymph. Furthermore, the endothelial cell wall of the lymph vessels may exert some selectivity in the uptake of solutes.

To summarise, there is no good evidence that proteins newly synthesised in the liver reach the blood circulation through lymphatics in other than minor amounts. The peculiarly permeable microvasculature of the liver allows the passage of much plasma protein into hepatic lymph and this accounts for the very high protein concentrations of hepatic lymph.

Other constituents of liver lymph

There is comparatively little information about constituents of hepatic lymph other than protein. Unlike intestinal or thoracic duct lymph, which is generally somewhat opalescent in the fasting state and turbid post-prandially, hepatic lymph is always clear, presumably reflecting differences in lipoprotein species and concentrations in the lymph. In rats, hepatic lymph cholesterol concentration is considerably less than that of plasma and is comparable with the concentration of cholesterol in the intestinal lymph of the fasted animal. Most of the difference between plasma and hepatic lymph cholesterol appears to be due to an excess of esterified cholesterol in the plasma (Friedman *et al.*, 1956). Studies on other species also indicate that hepatic lymph lipid concentrations are lower than those of plasma. Thus triglyceride, phospholipid and cholesterol concentrations in sheep hepatic lymph are roughly two-thirds those of plasma (Adams, 1964).

While the glucose content of the rat intestinal lymph is similar to that of plasma, hepatic lymph glucose concentration is roughly one and a third times plasma concentrations; urea concentrations in rat hepatic lymph and plasma are comparable (Friedman *et al.*, 1956). Small amounts of bile constituents are found in hepatic lymph in several species. Both conjugated and unconjugated bilirubin are present and it is tempting to speculate that unconjugated bilirubin reaches liver lymph by passing across the sinusoidal endothelial wall bound to albumin, while conjugated bilirubin of liver lymph may be derived by filtration from bile ducts into the portal tracts.

Cells and particulate material in liver lymph

Relatively large particles of foreign material have been noted to penetrate intestinal mucosa from the gut lumen and to enter both the venous system and the lymphatics (see Chapter 6). A few hours after intravenous injection of particles of titanium, 0.2–0.4 μm, or tantalum, 2–4 μm, these appear in liver lymph (Huggins & Froehlich, 1966; Dumont & Martelli, 1969). It is probable that they enter Kupffer cells and that the Kupffer cells free themselves from the sinusoid and pass into the space of Disse and thence to liver lymph (Smith, MacIntosh & Morris,

1970). They are then filtered by the hepatic lymph nodes and arrested there. Thus the reticulo-endothelial system of the liver while extracting foreign material from the blood by phagocytosis does not retain it in the hepatic parenchyma but in the related lymph nodes.

By contrast retrograde injection of particulate material into the obstructed common bile duct also results in passage of these particles into lymph, but these are not held up in the hepatic lymph nodes, instead going directly to the thoracic duct and blood circulation (Mallet-Guy *et al.*, 1962). In this instance the fact that the particles are not intracellular may explain why they are not filtered by hepatic lymph nodes.

Pathophysiology of liver lymph and haemodynamic disturbances

Hepatic venous obstruction

Experimental occlusion of the hepatic veins or supradiaphragmatic part of the inferior vena cava in animals results in a large increase in the flow and protein content of hepatic lymph and therefore of lymph in the thoracic duct. Similar changes in lymph flow result from chronic liver injury induced by carbon tetrachloride (Nix *et al.*, 1951). Hepatic venous obstruction is accompanied by dilatation of the space of Disse (Bolton & Barnard, 1931) and ascites develops rapidly (Bolton & Barnard, 1931; McKee, Schilling, Tishkoff & Hyatt, 1949; McKee, Wilt, Hyatt & Whipple, 1950; Nix *et al.*, 1951; Hyatt, Lawrence & Smith, 1955). This ascites appears to be due to transudation of fluid across the liver capsule. At laparotomy, droplets of fluid are seen forming on the surface of the swollen congested liver (Hyatt & Smith, 1954) and transplantation of the liver or a lobe of the liver to the thorax results in hydrothorax (Mallet-Guy, Devic, Feroldi & Desjaques, 1954; Aiello, Enquist, Ikezono & Levowitz, 1960). In acute hepatic vein occlusion in man (acute Budd–Chiari syndrome) marked ascites develops.

Increased lymph flow and ascites formation in hepatic venous obstruction probably arise from excess interstitial fluid formation as a result of raised intrasinusoidal pressure. Hydrostatic pressure in the sinusoid is in the range, 3–5 mmHg. In the isolated rat liver a small rise in hepatic venous pressure increases the liver volume and liver sodium, albumin and red cell spaces (Brauer *et al.*, 1959). In the cat, Greenway & Lautt (1970) have found that a two-phase increase in liver volume results from venous congestion and have proposed that the initial rise is a vascular congestion and that the second increase reflects increased interstitial fluid. As has been discussed already, the highly permeable nature of the sinusoidal endothelium, which allows proteins to pass rather freely into the extravascular space, suggests that effective oncotic pressure of hepatic blood plasma is low (Brauer, 1963; Greenway & Stark, 1971) and likely to be of little importance in controlling the formation of interstitial fluid (Brauer, 1963). Accordingly, relatively slight increases in hepatic venous pressure have marked effects on interstitial fluid and lymph formation in the liver.

Flow of thoracic duct lymph is increased by portal vein infusion of histamine (Nothacker & Brauer, 1950; Zeppa & Womack, 1963; Adesola & Schwartz, 1965). Histamine augments the flow and protein content of canine thoracic duct lymph suggesting an increase in the hepatic lymph contribution. An infusion of 5-hydroxytryptamine into the portal vein does not alter lymph flow. The explanation for the effect of histamine is not certain. Portal vein infusion of histamine in the dog reduces hepatic venous outflow and this is due to constriction of small hepatic veins (Andrews, Ritchie & Maigreith, 1973). Zeppa & Womack (1963), however, have proposed that histamine enhances permeability thus altering lymph production, but the hepatic sinusoid is highly permeable under normal conditions and an increase in sinusoidal pressure due to hepatic venoconstriction is a much more likely explanation.

Cirrhosis. Cirrhosis and two important consequences, ascites and portal hypertension, involve significant changes in flows in hepatic, intestinal and diaphragmatic lymph vessels and marked increases in flow and pressures in the thoracic duct. At laparotomy, patients with cirrhosis have greatly distended hepatic lymph vessels; increased numbers of such vessels in the hepato-duodenal ligament have been counted in histological preparations (Baggenstoss & Cain, 1957).

Ascites. The pathogenesis of ascites is complex and the factors involved include a profound disturbance of water and electrolyte metabolism, due in part to hyperaldosteronism, hypoalbuminaemia due to dilution by fluid retention and defective liver parenchymal function, and haemodynamic disturbances of the liver and splanchnic bed. The following discussion concentrates on these last two factors and on the role of lymph drainage of the liver, gastro-intestinal tract and peritoneal cavity in the development or prevention of ascites.

In health there is little free fluid in the peritoneal cavity and an accumulation of extracellular fluid in this compartment constitutes ascites. Such fluid as does pass into the peritoneal cavity in health is removed by peritoneal blood and lymph vessels. Diaphragmatic lymph vessels play a specially important part in draining the peritoneal cavity (Casley-Smith, 1964b) and particles, including red blood cells and foreign material such as India Ink, can be shown to enter these lymph vessels. Ascites arises where the input into the peritoneal cavity has outstripped the ability of these drainage mechanisms to clear the fluid.

In cirrhosis, liver lymphatics are overloaded. Injection of radio-opaque materials into human cirrhotic liver parenchyma outlines the lymphatics (Moreno *et al.*, 1963; Clain & McNulty, 1968; Dodd, 1970). It is probable that one factor in the pathogenesis of ascites in cirrhosis is similar to that already described in hepatic vein obstruction, that is, sinusoidal hypertension leading to an enormous increase in formation of hepatic interstitial fluid and lymph which transudes across the liver capsule. The extent to which hypoalbuminaemia with decreased plasma oncotic pressure contributes to ascites in cirrhosis is not clear but it is probable that the major factor is increased sinusoidal hydrostatic pressure. Filtration from the splanchnic circulation as a result of portal hypertension also contributes to the ascites of cirrhosis (see below) which can be regarded as a composite of variable proportions of hepatic lymph and splanchnic transudate.

Experimental ascites is most reliably produced by partial occlusion of the hepatic veins or inferior vena cava above the diaphragm. Acute extrahepatic portal hypertension from partial portal vein occlusion, with increased formation of interstitial fluid in the gut, will produce only a transient ascites (Volwiler, Grindlay & Bollman, 1950). Presumably, in this case, the capacity for reabsorption has not been exceeded. If, in addition, diaphragmatic lymph vessels are obliterated, ascites usually results, thus emphasising the importance of these drainage vessels (Raybuck, Weatherford & Allen, 1960). Fluid transport through diaphragmatic lymphatics may be responsible for the hydrothorax which frequently accompanies ascites.

Boyer, Maddrey, Basu & Iber (1969) have studied the importance of thoracic duct lymph in the transport of albumin from the peritoneal cavity in patients with portal hypertension and ascites. Labelled albumin added to ascitic fluid appeared relatively slowly in thoracic duct lymph but was found in lymph before it could be detected in plasma. Labelled albumin given intravenously appeared more rapidly in thoracic duct lymph and it was shown that a greater fraction of thoracic duct albumin was derived from plasma than from the ascites. It should however be remembered that the thoracic duct plays only a minor role in the lymph drainage of the peritoneal cavity. Most diaphragmatic lymph vessels communicate with lymph vessels running with the internal mammary blood vessels which drain to the right lymphatic duct.

Although experimental evidence summarised earlier suggests that liver lymph is of little importance as a route of transport of newly synthesised protein from the liver to the blood, it may be that in cirrhosis with ascites direct transfer of some of this protein from the hepatocyte to lymph and ascitic fluid occurs. Zimmon and his colleagues (1969) studied the appearance of intravenously injected ^{131}I-labelled albumin and ^{14}C-labelled albumin derived by hepatic biosynthesis from [^{14}C]carbonate in ascitic fluid in eight patients with alcoholic cirrhosis. Only 0.4–2.2% of the total plasma pool of albumin-^{131}I entered the ascitic fluid in two hours while 4.2–11.7% of the albumin-^{14}C appeared during the same period. In two patients in whom thoracic duct lymph was obtained 6.1 and 13.5% of newly synthesised ^{14}C-labelled albumin appeared in the lymph while 2.8 and 3.8% of ^{131}I-labelled plasma was recovered in the lymph. These results indicate a direct route from hepatocyte to lymph and ascitic fluid in patients with cirrhosis. It is not clear how this happens but it might be due to a relative impermeability of the sinusoid to protein delivered into the space of Disse as a result of deposition of basement membrane material in relation to the sinusoidal endothelium.

Portal hypertension. As a clinical entity chronic portal hypertension is commonly the result of cirrhosis, though occasionally it is due to extrahepatic obstruction of the portal vein. In either case, thoracic duct lymph flow is considerably increased. In cirrhotic patients hepatic lymph flow is greatly increased (Baggenstoss & Cain, 1957) while splanchnic bed hypertension results in a high flow of intestinal lymph. With extrahepatic portal obstruction, enhanced thoracic duct lymph flow is due only to an intestinal lymph contribution. Portal hypertension created experimentally by partial ligation of the portal vein in rats results in a striking dilation of small intestinal mucosal lymphatics, due, presumably, to increased formation of tissue fluid and lymph (Sotgiu, Cavalli & Gasbarrini, 1963).

The concentration of protein in intestinal lymph is considerably lower than that in hepatic lymph and this is ascribed to the relative permeabilities of the microvascular beds of the gut and liver (see Chapter 3). Since thoracic duct lymph protein concentration is largely determined by the contributions from the liver and gut the excessive production of hepatic lymph which occurs in cirrhosis raises thoracic duct lymph protein concentration. However, studies of patients with cirrhosis have shown that there is a group with advanced disease and marked portal hypertension in whom liver lymph and thoracic duct lymph protein concentrations are relatively low compared with the concentrations in cirrhotic patients where the main problem is a sinusoidal and post-sinusoidal venous obstruction (Dumont & Witte, 1966; Witte *et al.*, 1968; Dumont, Witte & Witte, 1975). In some of these patients sinusoidal permeability may have decreased as the result of the deposition of connective tissue resembling a basement membrane in the sinusoidal endothelium (Schaffner & Popper, 1963). Such a change in sinusoidal permeability occurring late in the course of the disease may reduce the ease with which a plasma filtrate rich in protein can escape from the blood in the face of obstruction to hepatic venous outflow, thus contributing to splanchnic vascular congestion and portal hypertension (Dumont *et al.*, 1975). Another explanation for low protein concentration in the *thoracic* duct lymph assumes that in these patients with severe chronic disease, an element of pre-sinusoidal obstruction has created marked splanchnic venous hypertension such that the gut makes a larger contribution to thoracic duct lymph than the liver. Successful porta-caval shunting by reducing portal pressure raises thoracic duct lymph protein concentration as once again the liver's contribution to the thoracic duct predominates (Witte, C. L., *et al.*, 1969).

Where the extrahepatic splanchnic bed is making a considerable contribution to thoracic duct lymph flow, any ascites which forms might seem to be mainly due to a rise in hydrostatic pressure in the splanchnic capillaries. However, hypoalbuminaemia due to hepatic disease is probably partly responsible since pure extrahepatic portal hypertension is not usually associated with chronic ascites. It is probable that in human cirrhotic portal hypertension there is an 'active' congestion of the splanchnic vascular bed due to venous obstruction and an increased arterial inflow to the gut (Witte, C. L., *et al.*, 1974). Studies of dogs with acute partial portal venous obstruction, increased splanchnic blood flow, or a combination of these, showed that the circulatory dynamics in the last group most closely correspond with those in human cirrhosis and it has been suggested that in this condition there is a failure of autoregulatory reduction in arterial inflow to the splanchnic area in response to venous outflow obstruction.

The observation by Lee & Duncan (1968) that when the pressure in the venous outflow from an isolated segment of jejunum is raised absorption of fluid from the lumen is directed towards the lymphatics, raises the interesting possibility that if this operates *in vivo*, patients with portal hypertension, for example in hepatic cirrhosis, would have an enhanced flow of

243

fluid and presumably solutes from gut lumen to lymph, allowing substances such as dietary nitrogenous compounds and drugs to enter intestinal lymph in amounts greater than usual and thus to bypass the liver where such substances are extracted.

Acute obstruction of the portal vein results in profound congestion of the intestinal mucosa. Extravasated blood in the infarcted tissue is partly removed by intestinal lymphatics (Barrowman, unpublished observations). In more limited vascular accidents of the gut, both venous and arterial occlusions, local lymphatic vessels probably clear the infarcted area of extravasated blood.

Porta-caval shunts and lymph flow

Reduction in splanchnic venous pressure in portal hypertension can be expected to alter intestinal and hepatic lymph flow. In dogs with ascites and portal hypertension produced by hepatic vein ligation, thoracic duct lymph flow was increased to thirteen times control values and thoracic duct pressure to three times control values. End-to-side porta-caval shunt in these animals reduced thoracic duct flow to five times control values and pressure to twice normal (Orloff *et al.*, 1967). On the basis that raised sinusoidal pressure is partly responsible for the augmented lymph flow, these results would suggest that end-to-side porta-caval anastomosis has not completely decompressed the liver. Side-to-side anastomosis, however, essentially returned thoracic duct flow and pressure to normal and effective hepatic decompression in this instance can be explained by retrograde blood flow in the valveless portal vein. Parallel results are obtained in man where porta-caval shunts are used to decompress the splanchnic bed and hepatic sinusoids in cirrhosis (Witte *et al.*, 1969b). End-to-side porta-caval shunt reduces thoracic duct lymph flow with a rise in lymph protein concentration. This is explained by a reduction in the protein-poor intestinal lymph contribution to thoracic duct lymph with relatively more protein-richer liver lymph contributing. In a patient with a thrombosed porta-caval anastomosis, thoracic duct lymph flow rose and its protein concentration fell, suggesting a renewed large contribution from the

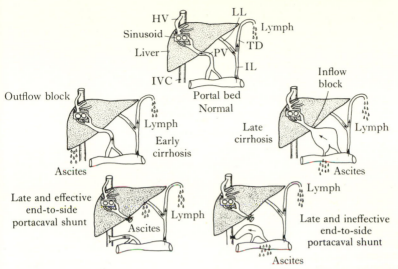

Fig. 8.4. The formation of ascites in hepatic cirrhosis. In early cirrhosis, ascitic fluid is mainly derived from the liver due to a degree of post-sinusoidal obstruction. In late cirrhosis ascites, an element of pre-sinusoidal obstruction leads to increased formation of tissue fluid in extrahepatic tissues. With an effective porta-caval shunt, ascites is reduced but if the shunt blocks, ascites collects as a result of extrahepatic portal congestion. Abbreviations: TD, thoracic duct; PV, portal vein; HV, hepatic vein; LL, liver lymph; IVC, inferior vena cava; IL, intestinal lymph. (Witte *et al.*, 1969.)

splanchnic bed. Fig. 8.4 is a schematic representation of the mechanisms thought to operate in determining lymph flow in cirrhosis with portal hypertension with or without porta-caval shunt.

Presinusoidal portal hypertension

Schistosomal hepatic fibrosis. While the obstruction to portal blood flow in hepatic cirrhosis is considered to be mainly sinusoidal and post-sinusoidal, schistosomal hepatic fibrosis principally causes a pre-sinusoidal obstruction (Ramos, Saad & Leser, 1964). The consequent portal hypertension might be expected to produce an increased capillary filtration in the splanchnic area, with enhanced lymph flow. If ascites is present this would also add to the load on lymph channels, in this case those draining the peritoneal cavity, particularly the diaphragmatic lymphatics. Since ascites frequently does occur in this condition it would appear that in addition to portal hypertension other factors are operating, as uncomplicated portal hypertension does not generally result in ascites. Sadek *et al.* (1970) have studied the thoracic duct *post mortem* in twenty patients with this condition and found dilatation and tortuosity of the vessel compared with controls. These abnormalities were most marked in patients with severe ascites. The capacity of the dissected thoracic duct in some cases was found to be markedly increased. Lymphangiography and direct inspection of the thoracic duct in other patients showed dilatation and tortuosity similar to that seen in post-mortem material and pressure in the thoracic duct was reported to be increased twice to four times normal. Normal values, however, were not given and in view of the variation in pressure measurements in lymphatics (see Chapter 3) this statement must be viewed with care. Blood staining of thoracic duct lymph was frequently noted in this condition. This study does not show that schisto-somal hepatic fibrosis produces an enhanced lymph flow from the territory of the portal vein alone and the presence of ascites suggests that liver lymphatics may be overloaded as well.

In a subsequent study, however, these same authors found dilated splenic lymph vessels, some containing red blood cells in this condition. Splenectomy was found to reduce thoracic duct lymph flow and pressure though these measurements did not correlate well with the fall in portal venous pressure which also resulted from splenectomy.

Thus there is some indirect evidence that in this condition enhanced capillary filtration from the splanchnic bed due to portal hypertension may result in excess lymph formation. The presence of red cells in lymph may be due to the opening of channels of communication between veins and lymphatics in the portal bed, or to extravasation of red cells from congested capillary vessels. There is probably also an enhanced flow of liver lymph in this condition as dilated subcapsular hepatic lymph vessels were noted by Aboul-Enein (1973) in post-mortem examination of patients with schistosomal hepatic fibrosis and ascites. In a recent study, Ismail & Aboul-Enein (1976) found that the total protein content of thoracic duct lymph in schistosomal hepatic fibrosis is lower than that in control subjects (mean values 35 and 57% of total plasma protein concentration, respectively) suggesting that the main source of the lymph and ascites in these patients is the splanchnic bed. Obviously, the haemodynamic changes in this condition are complex with both pre- and post-sinusoidal obstruction playing a part. Some clarification of this problem has come from a study reported by El-Zawahry *et al.* (1977). In a large series of patients with the disease it was found that thoracic duct lymph protein concentration was 66% that of serum in those patients who did not have ascites and only 29% in those who had tense ascites. The higher protein concentration in thoracic duct lymph in the early stage of the disease could be explained as due to combined pre- and post-sinusoidal obstruction while the lower protein concentration in the group with advanced disease probably reflects a large contribution from the splanchnic bed. It appears then, that in this condition as in cirrhosis the proportions of lymph derived from splanchnic and hepatic sinusoidal beds are variable and are related to the stage of evolution of the disease.

Idiopathic portal hypertension. In a group of patients with portal hypertension but who lack histological evidence of cirrhosis, 'idiopathic portal hypertension', Samanta *et al.* (1974) have found flow rates of thoracic duct lymph to be approximately three times control values. The duct was distended and lymph pressure was increased. In a few patients, dilated liver lymph vessels were demonstrated radiologically. The thoracic duct lymph:serum protein ratio in this group was 0.29 while that in a group of patients with cirrhosis was 0.41, both surprisingly low values; it was suggested that the contribution of intestinal lymph to the thoracic duct may be more important in this condition that in cirrhosis. This would be in keeping with the suggestion that portal hypertension in these patients is predominately pre-sinusoidal (Datta, 1969).

Thoracic duct drainage in the management of ascites and portal hypertension
In cirrhotic patients with ascites, the thoracic duct is dilated to twice to four times its normal size and the rate of flow of lymph is increased six to ten times over normal values (Dumont

& Mulholland, 1960). Since ascites is considered to be due to enhanced extracellular fluid formation in the liver, associated with an augmented liver lymph flow, it has been suggested that the thoracic duct is not able to clear the large amounts of lymph presented to it, possibly because of a functional obstruction at the jugulo-subclavian tap in the neck (see, for example, Dumont, 1975*a*, *b*). Since raised systemic venous pressure is not common in cirrhosis, such functional obstruction is likely to arise at the junction itself or to reflect diminished propulsion of lymph from the thoracic duct. If the resistance is located at the junction, surgical techniques to bypass this resistance, or thoracic duct drainage with appropriate intravenous replacement of fluid, electrolytes and protein, might be expected to diminish ascites by improving the efficiency of lymph drainage though studies of experimental ascites in dogs with hepatic vein ligation treated by external thoracic duct drainage have suggested that this manoeuvre is only of limited effectiveness in decompressing the obstructed liver and overburdened liver lymphatics (Orloff *et al.*, 1966).

The thoracic duct–jugular venous junction
There are a number of anatomical variations in this structure, even within a species. For example in about 50% of human subjects an ampullary dilatation of the duct occurs shortly before the jugulo-subclavian tap. Often, multiple small lymph trunks arise from the ampullary dilatation and discharge into the jugular or subclavian veins. There is a valve in the vein wall which has paired, unusually long, leaflets and is set obliquely. About 1 cm back into the thoracic duct two valves are found which also assist in preventing venous reflux into the lymphatic vessel. Detailed descriptions of the anatomy of the region have been given by Pflug & Calnan (1956), Zemel & Gutelius (1965) and Bradham & Takaro (1968). The last authors noted that in both man and dog the lymphatic vessel tapers as it approaches the vein. There is a pressure gradient across the anastomosis but, due to different techniques of measurement, there are wide differences in reported values. In general it appears that the gradient, lymphatic to vein, is of the order of 3 or 4 mmHg (Dumont, 1975*b*).

Lymph enters the blood in a phasic fashion governed by respiratory effects on the thoracic duct and on pressure in the neck veins. Inspiratory effort increases abdominal pressure and directs lymph towards the low pressure area of the thorax, where valves assist in directing the lymph cephalad. Expiration, by raising intrathoracic pressure, directs lymph towards the neck. Lymph is held up here as the lymphatico-venous valve is closed during expiration, while the vein wall is stretched with blood. The subsequent inspiration collapses the vein wall and the lymph enters the vein. The evidence for this sequence of events can be found in pressure measurements in dogs (Battezzati & Donini, 1972) and lymphangiographic studies in dogs and man (Kinmonth & Taylor, 1956; Browse, Rutt, Sizeland & Taylor, 1974). The contribution of intrinsic contractility of lymphatic vessels to lymph propulsion is discussed in Chapter 3.

The concept that the lymphatico-venous junction in the neck can constitute a serious obstruction to lymph flow in conditions where large amounts of lymph are being transported to the blood, can be considered in cases of congestive cardiac failure, where right-sided heart failure leads not only to increased venous pressure in the large veins of the neck but also to increased lymph formation, and in some cases, ascites (Witte *et al.*, 1969a). Data on this subject are conflicting though surgical anastomosis of a dilated thoracic duct to low pressure intrathoracic veins will relieve ascites in dogs with experimental right heart failure (Cole *et al.*, 1967).

As already discussed suprahepatic venous obstruction leads to ascites, greatly elevated liver lymph and thoracic duct lymph flows and marked dilatation of the duct in dogs. Where blood flow in the liver is obstructed at a sinusoidal (cirrhosis) or pre-sinusoidal (schistosomal hepatic fibrosis) level, lymph flows are also elevated, due to excess hepatic and splanchnic bed lymph formation. In both experimental hepatic venous obstruction and cirrhosis, re-anastomosis of the distended lymph vessel to another neck vein reduces thoracic duct pressure and ascites and results in a diuresis and weight loss (Kottakis & Agapitides, 1970; Donini & Bresadola, 1974; Serenyi, Magyar & Szabó, 1976). In the study by Donini &

Bresadola, lymphangiography demonstrated a reduction in the tortuosity of the thoracic duct of cirrhotic patients with a re-fashioned thoracic duct–jugular vein tap, and splenic pulp pressure, an index of portal vein pressure, fell in four of five patients so examined from 21 to 15 mmHg (mean values). Such results suggest that the normal jugulo-subclavian tap is unable to accommodate greatly enhanced lymph flows. Why this should be, however, is a matter of speculation. Clearly there is a limited period of time when lymph enters the bloodstream, that is, in the early phase of inspiration. In addition it has been suggested that distension of the thoracic duct leads to a decrease in the compliance of that vessel with decreased flow across the junction (El-Gendi & Zaky, 1970). A number of studies have attempted to analyse the resistance of the jugulo-subclavian tap by measuring flow and the lymphatic–venous pressure gradient (see review by Dumont, 1975*b*). No general agreement has yet emerged as to the physiological explanation for the apparent obstruction. Warren, Fomon & Leite (1968) have shown that in cirrhotic patients the thoracic duct could be completely decompressed through a cannula of internal diameter 1.1 mm, considered to be smaller than the actual junction, and thus has challenged the view of a resistance at the junction. As Dumont (1975*b*) points out, the ostium does not appear to take part in the dilatation of the thoracic duct seen in cirrhosis, and precise measurements of its diameter are needed. It is important to recall, too, that apart from structural considerations, the phasic nature of lymphatic–venous flow may contribute to the imped-ance of the discharge of lymph and thus simple decompres-sion by cannulation cannot be compared with the intact system.

Since thoracic duct drainage reduces the pressure in the portal venous bed in portal hypertension, as measured by splenic pulp pressure (Dumont & Mulholland, 1962), there has been a little interest in the use of this procedure in the management of oesophageal variceal haemorrhage (Bowers, McKinnon, Marino & Culverwell, 1964; Cueto & Currie, 1967; Datta *et al.*, 1971). Evaluation of this treatment is difficult and it is not widely used. It is interesting that in this condition thoracic duct lymph is haemorrhagic and Dumont, Witte, Witte & Cole (1970) showed that neither intestinal nor hepatic lymph contained blood. It is probable that the blood enters the lymph through lymphatico-venous communications between oesophageal veins and the thoracic duct (Ludwig, Linhart & Baggenstoss, 1966).

The transport of bile constituents in liver lymph in cholestasis

The possibility that in obstructive jaundice, lymph carries bile to the systemic circulation has been considered for many years. As early as 1795 Saunders suggested that, in obstructive jaundice, bilirubin enters the blood through liver lymph. In 1874 Fleishl found that when the bile duct was ligated, bile acids and bilirubin appeared earlier and in higher concentration in the hepatic lymph than in the plasma. Since then experimental studies have shown that this lymphatic route does exist, but its quantitative importance is still unclear. Obstruction of the common bile duct in several species leads to a marked rise in the flow of hepatic lymph, shown for example in rats by Friedman *et al.* (1956) and in anaesthetised rabbits by del Rio Lozano & Andrews (1966). In the study by Friedman *et al.* (1956) increased lymph concentrations of bile acids and bilirubin were also found. Retrograde injection of India Ink into the common bile duct did not result in the appearance of carbon particles in liver lymph, suggesting that no rupture of fine intrahepatic radicles had occurred. Burgener & Adams (1976), however, have shown that radiographic contrast medium infused retrogradely into the common bile duct of dogs, at a pressure slightly exceeding biliary pressure, opacified the deep hepatic lymphatics. No anastomoses were found between the deep and superficial hepatic lymph system.

Bile pigments. Icteric lymph is found in obstructive jaundice but it has been suggested that lymph is only important in the transport of bile constituents in the first few hours after obstruction (Whipple & King, 1911; Bloom, 1923; Mayo & Greene, 1929; Shafiroff, Doubilet & Ruggiero, 1939; Gonzalez-Oddone, 1946; Hanzon, 1952; Ritchie, Grindlay & Bollman, 1956; Carlsten, Edlund & Thulesius, 1961; Dumont, Doubilet,

Witte & Mulholland, 1961; Mallet-Guy, Michoulier & Baev, 1965; Witte, Dumont, Levine & Cole, 1968). Recently, attempts which will now be described have been made to establish whether this is so.

On the basis that all bilirubin circulating in the body fluids is protein-bound, Dumont (1973) measured concentrations of bilirubin and protein in thoracic duct lymph and serum in patients with obstructive jaundice and used lymph:serum protein and bilirubin ratios in order to determine the source of bile pigment in lymph. In a series of sixteen patients with obstructive jaundice, lymph:serum ratios for protein and bilirubin were similar, suggesting that in those patients bile enters the lymph by filtration from the blood in the liver, the extrahepatic portal bed and other tissues drained by the thoracic duct. In five patients, the results suggested that extra bilirubin entered the hepatic lymph, coming presumably from hepatocytes and possibly fine biliary canaliculi, while in two patients lymph:serum bilirubin ratios were less than those of protein, suggesting some clearance of bile pigment from hepatic lymph by the liver. In eleven of fourteen of these patients in whom thoracic duct lymph was diverted, serum bilirubin levels fell, indicating that a considerable amount of bilirubin passes through the thoracic duct daily, and where the lymph:serum ratio was comparable with that of protein, this would mean that the extravascular circulation of bilirubin is of comparable degree to that of proteins; probably at least 50% of the circulating bilirubin pool passes through the thoracic duct daily. These observations relate to established chronic biliary obstruction.

Experimental studies by Szabó, Magyar & Jakab (1975a) seem to suggest a more important role for hepatic lymph in the transport of biliary solutes in acute biliary obstruction. When $Na^{125}I$ and ^{131}I-labelled albumin were infused under pressure into the common bile duct of dogs it was found that at high pressures (40 mmHg) the amounts of iodide and labelled albumin in circulating blood plasma were comparable, but at lower pressures higher concentrations of iodide than albumin were found in plasma. Thoracic duct lymph concentrations of labelled albumin were higher than those of plasma.

Their result can be interpreted as showing that at lower pressure the leak of both tracer substances from small bile ducts into the interstitial space of the portal tract leads to fluid and small solute absorption by peribiliary capillary vessels, with consequent increase in protein concentration in the interstitial fluid, leading to high concentration of the labelled albumin in hepatic lymph. The removal of the albumin from the hepatic interstitial space is slower than for the iodide, thus creating a disparity in their concentrations in peripheral blood. At higher pressure, however, bile probably leaks from biliary canaliculi into the space of Disse, and both solutes enter the blood circulation in roughly equal proportion as a result of the high permeability of the sinusoidal endothelium.

In a further series of experiments this same group showed that occlusion of the common bile duct of dogs leads to a rapid rise in bilirubin concentration in thoracic duct lymph, the maximum concentration being reached in four to six hours after obstruction. Lymph concentrations exceeded plasma concentrations throughout the entire period of observation (twenty-four hours after obstruction). Furthermore it could be calculated that the amount of bilirubin transported by the lymph exceeded that transported to the circulation by the hepatic veins, after the first two hours following occlusion of the common bile duct.

Thus the results of these experiments by Dumont (1973) and Szabó, Magyar & Jakab (1975a), although studying different species under different conditions, are in keeping with the suggestion that, in the early period of bile duct obstruction, lymph constitutes the major route of bile pigment transport to the circulation, but the bile pigment in lymph in chronic biliary obstruction is chiefly derived from the plasma by filtration. In the latter case, an equilibrium has been established where raised biliary pressure has decreased or stopped biliary secretion and the excess bile pigment now merely circulates through blood, interstitial fluid and lymphatics, in association with proteins.

The difference in transport of conjugated bilirubin in acute and chronic cholestasis in cholecystectomised dogs has been confirmed by Bergan, Taksdal & Sander (1975). Thus

Liver lymph

[^{14}C]bilirubin given by intravenous injection to these animals one hour after induction of cholestasis was rapidly conjugated and reached the thoracic duct in high concentration, and no evidence was found for delivery of conjugated bilirubin directly to the blood; during chronic cholestasis the bilirubin conjugates were returned directly to the blood after conjugation in the liver. In a further study in dogs with thoracic duct fistulae and acute biliary obstruction Bergan, Taksdal & Enge (1975) obtained contrasting results finding that *non-erythroid* bilirubin, synthesised *in vivo* from [^{14}C]δ-aminolaevulinic acid, is delivered in part into the blood circulation directly via sinusoidal blood; in chronic cholestasis the *non-erythroid* bilirubin in lymph appears to be secondarily derived from circulating blood by filtration.

Bile salts. Under normal circumstances, more than 99.7% of bile salts are efficiently contained within the entero-portal-biliary system. As a consequence, the concentration of bile salts in the portal venous blood is much higher than in systemic blood, where only trivial amounts are generally found (see Heaton, 1972). This is due to the great capacity of the hepatocyte to clear portal venous blood of bile salts. However, with biliary obstruction and increased pressure in the common bile duct, the secretion of bile salts is reduced and their synthesis in the hepatocyte is ultimately halted (Strasberg, Dorn, Small & Egdahl, 1971). In acute biliary obstruction, lymph becomes a route for the transport of bile salts to the systemic circulation. Ligation of the common bile duct in dogs causes a large rise in the concentration of bile salts in thoracic duct lymph, and their transport by this route was found to rise from negligible amounts (0.067±0.042 nmoles per min per kg body weight) to a maximum between six and eight hours after ligation (4.06±0.081 nmoles per min per kg body weight) at which point, thoracic duct lymph flow had risen by approximately 60% over its basal rate (Szabó, Magyar, Szentirmai, Jakab & Mihaly, 1975). The fall in lymphatic transport thereafter is presumably related to suppression of bile acid synthesis in the hepatocyte by the obstruction. In contrast to the lymphatic transport, maximum venous blood transport appeared to

occur in the first two to four hours after ligation. From plasma
and lymph concentrations of bile salts and albumin, it could
be determined that after two to four hours of biliary obstruction, the amount of bile salt transported by lymphatics exceeded that reaching the blood directly, that is, via the hepatic
veins.

Bile salts in plasma and lymph are bound to albumin but
are not protein-bound in bile, where they exist in micellar
aggregates which also contain phospholipids and cholesterol.
Therefore, in biliary obstruction, it may be that bile salts
escaping either from biliary canaliculi into the space of Disse
or from biliary ducts into periportal connective tissue, are
rapidly picked up by the relatively large amounts of albumin
in hepatic interstitial fluid and transported by both lymph and
blood. From the systemic blood, bile salts are probably
recycled through the interstitial space and back via the lymph.
Dumont (1973) has noted that thoracic duct drainage in patients
with obstructive jaundice reduced pruritus within a few hours.
If it is accepted that the pruritus of cholestasis is caused by
a rise in the concentration of bile salts in systemic blood, this
observation would point to a significant delivery of bile salts
to the circulation from the liver by lymph.

The relatively important part played by hepatic lymph in the
transport of bile constituents in acute biliary obstruction
suggests that it is the passage of bile into the interstitial space
of the portal tract which is important since bile escaping into
the space of Disse from canaliculi would be expected to pass
in large quantity into sinusoidal blood.

The lymphatic system is intimately involved in a wide variety of pathological processes in the alimentary tract. For example, it plays an important part in the dissemination of gastro-intestinal neoplasms and in certain parenchymatous diseases of the liver and pancreas it acts as a route of drainage for large quantities of fluids and solutes (see Chapters 5 and 8). This chapter considers only those disorders of function which arise as a direct consequence of pathological processes in the gastro-intestinal lymphatic system.

Intestinal lymphangiectasia

Dilatation of intestinal lymph vessels may be either primary, that is, a congenital defect of development of the lymphatic vessels of the alimentary tract, or secondary to a pathological process in the gut or route of its lymphatic drainage.

Primary intestinal lymphangiectasia

Intestinal lymphangiectasia was described in man in 1961 by Waldmann and his colleagues. This uncommon condition, which usually presents by early adult life, is characterised by a protein-losing enteropathy with hypoalbuminaemia and structural abnormalities of the lymphatic vessels of the small intestine and of the mesentery. A variable degree of steator-rhoea, hypocalcaemia, hypomagnesaemia and lymphopenia is frequently present and chylous ascites, chylothorax and chyluria are commonly present. Normal xylose absorption distinguishes this condition from other malabsorptive states. A condition similar to human intestinal lymphangiectasia has been described in the dog (Campbell, Brobst & Bisgard, 1968).

In the description which Waldmann *et al.* (1961) gave of intestinal lymphangiectasia, the most striking structural change in lymphatics was a dilatation and telangiectasia of mucosal and submucosal lymph vessels. Associated with this was the presence of dilated lymph vessels in the serosa and mesentery. Lipid-laden macrophages are frequently observed in the lumen of the vessels and the affected intestine is oedematous. Plate 9.1 illustrates the appearances of the mucosa of the small intestine in this condition. In some patients the changes occur patchily in the intestine.

Interpretation of the appearance of lymph vessels in histological preparations of small intestinal biopsies can be difficult. Spaces which can be mistaken for lymph vessels in the centre of the villus can be created by faults in sampling or in the histological preparative procedure. Bank, Fisher, Marks & Groll (1967) have made an extensive study of small bowel mucosal biopsies and consider that pathological intestinal lymphatic dilatation is present when more than four of fifty intestinal villi show lymphatic spaces, or when the dilated lacteals are greater than 129 μm in length or 64 μm in width.

While most descriptions of this condition indicate that the small intestine is affected, isolated case reports have described this type of change in the large intestine (Plate 9.2; Schaefer, Griffen & Dubilier, 1968; Ivey, den Besten, Kent & Clifton, 1969; Griffen *et al.*, 1972). In the report by Ivey *et al.*, the condition apparently was limited to the large intestine and was associated with a protein-losing enteropathy, steatorrhoea and lymphopenia. The presence of steatorrhoea suggests small intestinal involvement, though this could not be demonstrated.

Dobbins (1966*a, b*) examined the ultrastructure of the intestinal mucosa of a patient with this condition making a comparison with normal mucosa. The principal findings were the presence of very prominent intracellular 'fibrils' in the lymphatic endothelium, possibly a reflection of chronically raised intralymphatic pressure. Structures closely related to the vessel (the basement membrane, collagen fibres, smooth muscle and connective tissue cells) were all unusually prominent. Large

257

lipid droplets were found at the base of absorptive cells and in the lymphatic endothelium, and numerous chylomicra were found both within the lumen of dilated vessels and in the extracellular space of the villi. These observations are compatible with a block in lymphatic flow and probably an associated longstanding hypertension within the vessels. Mistilis, Skyring & Stephen (1965) have observed changes in the mucosal epithelium of the small intestine in intestinal lymphangiectasia. Such changes include an attenuation of the epithelial cells, with displacement of their nuclei towards the apices of the cells.

The frequent association of morphological abnormalities in other parts of the lymphatic system with intestinal lymphangiectasia suggests that a primary developmental disorder of lymphatic vessels accounts for a large proportion of cases. Such association has been shown by Pomerantz & Waldmann (1963) and Mistilis *et al.* (1965). In several cases, chylous effusions are present in the peritoneal or pleural cavities. Garciá-Alvarez *et al.* (1974) have described a patient with intestinal lymphangiectasia and congenital lymphoedema of some limbs who developed acute hepatitis. The jaundice disappeared more slowly from the lymphoedematous limbs presumably because albumin-bound unconjugated bilirubin lingered in the static interstitial fluid. In a group of cases where intestinal lymphangiectasia is acquired, obstructive causes, such as retroperitoneal fibrosis, chronic pancreatitis or disease of mesenteric lymph nodes may be responsible. In occasional cases an isolated structural abnormality is found in a major lymph trunk. Molnár, Karoliny, Mozsik & Németh (1974), reported a case in which the patient had an obstruction of uncertain cause, possibly of congenital origin, at the level of the cisterna chyli and the back pressure created led to a protein-losing gastroenteropathy and a reversal of the flow of lymph in the liver, whereby the periportal lymph which normally drains to the hilar lymph vessels and thence to the cisterna chyli flowed to the liver capsule and thence by lymph vessels through the diaphragm to parasternal lymphatics.

The symptoms and signs associated with intestinal lymphangiectasia are explicable on the basis of either rupture of dilated intestinal lymph vessels with chylous discharge into the gut lumen or transudation of lymph and tissue fluid from the oedematous lamina propria of the gut. Either process would be expected to allow escape of a protein- and lipid-rich fluid. On duodenal intubation, a chylous fluid having a protein content of up to 0.5 g per 100 ml was aspirated in one such patient. It was observed that this fluid contained considerable amounts of long-chain fatty acid (Stoelinga, van Munster & Slooff, 1963). Further evidence for a chylous leak into the gut lumen comes from the demonstration of radiological contrast medium outlining small intestinal mucosal folds following lymphangiography in some patients with intestinal lymphangiectasia (Gold & Youker, 1973). Protein loss leads to hypoalbuminaemia and to oedema. It is possible that nutritional deficiency may aggravate the hypoproteinaemia. Oedema in these patients is frequently asymmetrical, suggesting that other local anomalies of lymphatic drainage are also present. Steatorrhoea is of variable degree; in one patient studied by Mistilis *et al.* (1965), faecal fat was estimated at 8.5 g per day on a fat-free diet. These authors regard the loss of protein and fat as an exudation of a chylous fluid across the mucosa. Lymphocytopenia is frequently present, and serum immunoglobulins are often reduced. Strober, Wochner, Carbone & Waldmann (1967) have studied patients with intestinal lymphangiectasia with special reference to immunological deficiency. These patients showed a striking failure to reject skin homografts and the authors concluded that the immunologic deficiency results from a loss of both lymphocytes and immunoglobulins into the gastro-intestinal tract. McGuigan, Purkerson, Trudeau & Peterson (1968) have shown that the immunologic defects in this condition are correlated with increased loss of immunoglobulins, rather than an associated defect in immunoglobulin production, since the immunoglobulin-bearing cells of the lamina propria of the small intestine in this condition are present in normal numbers. Intestinal lymphangiectasia is, therefore, an important form of protein-losing gastroenteropathy.

Protein-losing gastroenteropathy

The leakage of serum albumin into the normal gastro-intestinal tract has been demonstrated by Birke, Liljedahl, Plantin & Wetterfors (1959), Gullberg & Olhagen (1959), and Holman, Nickel & Sleisinger (1959). The extent of catabolism of serum proteins in the gastro-intestinal tract under normal circumstances has been the subject of considerable controversy (see the review by Waldmann, 1966). A wide spectrum of plasma proteins appear in small quantities in the gastro-intestinal secretions of normal human subjects. In the normal situation, there is reabsorption of a proportion of the amino acid liberated by such catabolism. Hypoproteinaemia results when catabolism exceeds the synthetic capacity of the liver, even if the absorption of the amino acids liberated by the catabolic process is almost complete. Excessive plasma protein entry into the human gastro-intestinal tract in alimentary disease has been recognised for some time. Citrin, Sterling & Halsted (1957) demonstrated the presence of albumin in gastric secretion in a patient with giant rugal hypertrophy of the stomach and since then protein-losing gastroenteropathy has been described in a wide variety of pathological disorders, which include neoplastic and inflammatory lesions. A number of cases of previously unknown origin have recently been identified as being due to abnormalities of the intestinal lymphatic vessels. Waldmann (1966) has reviewed the pathological conditions which have been shown to be associated with protein-losing enteropathy. In addition to protein loss, isotope and chemical techniques have demonstrated that iron (Ulstrom & Krivit, 1960), calcium (Milhaud & Vesin, 1961) and lipids (Mistilis et al., 1965) are also lost into the gut lumen in many cases of protein-losing enteropathy.

Various techniques have been used to demonstrate abnormal loss of plasma protein into the gastro-intestinal tract. These techniques include turnover studies of albumin labelled with ^{131}I or [51]Cr. Caeruloplasmin labelled with [67]Cu and ^{131}I-labelled polyvinylpyrrolidone have also been used for this purpose and excretion of ^{51}Cr-labelled transferrin is routinely used to detect protein-losing enteropathy in clinical practice. In this method the radio label is given as inorganic chromium which attaches to transferrin *in vivo*.

Protein loss in intestinal lymphangiectasia

The extensive loss of large protein molecules into the gastro-intestinal tract in intestinal lymphangiectasia reveals that, as in other protein-losing gastroenteropathies, the molecular size of proteins plays little role in determining the extent to which they are lost (Waldmann & Schwab, 1965); that is, there seems to be no molecular sieving. This is a little surprising since most lymph protein is derived from blood plasma by molecular sieving at the capillary endothelium (see Chapter 3). However, it is important to remember that in intestinal lymphangiectasia many of the large protein molecules which are lost, such as lipoproteins and certain immunoglobulins, are derived from the intestinal mucosa itself.

While the effects of intestinal lymphangiectasia may be largely explained by structural and functional abnormalities of intestinal lymph vessels, lymphatic abnormalities of other regions may contribute to the clinical picture. These include chylous ascites and lymphoedema of limbs. The occurrence of chylous ascites in this condition may be due to rupture of dilated mesenteric lymph vessels (Vescia & Davis, 1965) though abnormalities of lymph vessels draining the peritoneal cavity may also be responsible. Mizuno *et al.* (1966) have proposed the interesting hypothesis that where the pathological process extends to involve the lymphatics draining the pancreas, necrotic changes in the gland may lead to an insufficiency of pancreatic exocrine secretion with consequent failure of digestion and absorption, thus aggravating protein and fat deficits.

Although primary intestinal lymphangiectasia is generally considered a rare condition, it may be more frequent than has previously been thought (Roberts & Douglas, 1975; Vardy, Lebenthal & Schwachmann, 1975). The diagnosis has to be considered in unexplained protein-losing enteropathy with steatorrhoea and small intestinal biopsy may need to be performed repeatedly to establish the diagnosis.

Pathophysiology of gastro-intestinal lymphatics

Secondary intestinal lymphangiectasia

Many diseases produce an obstruction to lymph flow from the intestine. Dilatation of lymphatic vessels with obstruction of lymph drainage of the small intestinal mucosa is seen in acute and chronic inflammatory disease of the small intestine and its mesentery. These changes are also frequently seen in neoplastic disease of mesenteric lymph nodes, while chronic pancreatitis and retroperitoneal fibrosis may also obstruct mesenteric lymph flow (for review, see Frank & Kern, 1968). Obstruction and consequent dilatation of terminal intestinal lymphatic vessels, may occur in Whipple's disease (see below) and a patient has been described by Dobbins (1968) who suffered from hypo-β-lipoproteinaemia and a coexistent intestinal lymphangiectasia of moderate degree. Dobbins has proposed that the lymphangiectasia in this case was due to lymph flow obstruction caused by an infiltration of mesenteric lymph nodes by fat filled macrophages, a histological appearance demonstrated by lymph node biopsy. These various conditions can be regarded as causing secondary intestinal lymphangiectasia. McMahon & Neale (1969) described a case of protein-losing gastroenteropathy with lymphangiectasia which was probably due to filarial obstruction of mesenteric lymph vessels. Chronic constrictive pericarditis is also frequently associated with lymphangiectatic changes in the gastro-intestinal tract. The pathogenesis of these changes in heart disease deserves particular comment (see below).

Management of intestinal lymphangiectasia

The clinical management of intestinal lymphangiectasia presents an interesting opportunity for rational use of data obtained from experimental work on lymphatic physiology in animals. In 1953, Borgström & Laurell demonstrated a large rise in thoracic duct lymph flow of relatively high protein content in rats given corn oil by intragastric tube. Peak flow was reached some two hours before maximum lymph lipid levels were obtained. Similar observations have been made by Simmonds (1954). Such studies indicate that the digestion and absorption of a fat meal is accompanied by an increased lymphatic transport of protein over several hours. Since flow rates increase, it is likely that lymphatic pressures rise, and this would be particularly pronounced if there was an element of obstruction. The origin of the increased fluid and protein content of intestinal lymph during fat absorption is probably the blood plasma, and the changes in lymph are probably due to alterations in intestinal blood flow (see Chapter 7). A fat meal should, therefore, be expected to increase the tendency for protein transudation in intestinal lymphangiectasia; a low fat diet has been shown by Jeffries, Chapman & Sleisinger (1964) to increase serum albumin levels and reduce albumin catabolism as judged by the turnover of ^{131}I-labelled albumin (Fig. 9.1). Furthermore, since long-chain fatty acids (fourteen carbon atoms and longer) are principally transported in intestinal lymph as resynthesised triglyceride after they have been absorbed and shorter chain fatty acids pass unesterified into the portal vein (Bloom, Chaikoff & Reinhardt, 1951; see Chapter 7), a diet consisting of medium-chain triglyceride (having fatty acids of six to twelve carbon atoms) will divert fat transport from the lymphatic route. Holt (1964) found that such a diet greatly reduced gastro-intestinal protein loss in a child with intestinal lymphangiectasia. It might be concluded that in this condition, a deleterious effect is produced by feeding long-chain triglyceride which directly and indirectly increases the load of fluid, protein and lipid on an already inadequate system. Other studies however, suggest that feeding medium-chain fats does not reduce the formation of lymph in patients with disease of the lymphatic vessels. For example, Frank, Kern, Franks & Urban (1966) found that in a patient with chylous ascites due to lymphatic obstruction by lymphosarcoma and fibrosis, medium-chain fats did not reduce the rate of accumulation of ascites though its lipid content was markedly reduced. The observation by Holt (1964) that medium-chain fat feeding reduces gastro-intestinal protein loss may have a complex explanation since it has been noted that short-chain fat (tributyrin) absorption is accompanied by augmented intestinal lymph flow (Simmonds, 1955) and instillation of medium-chain triglyceride into the duodenum of rats also increases lymph flow with an increase in the amount of protein carried in that lymph per unit time (Barrowman & Turner, 1976). It

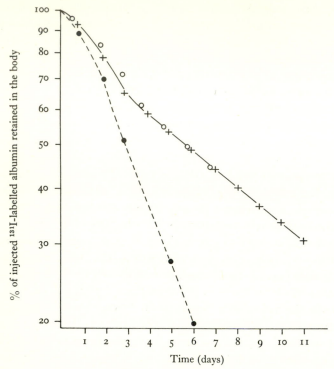

Fig. 9.1. Turnover of ^{131}I-labelled albumin in a patient with intestinal lymphangi-ectasia. The percentage of injected albumin retained in the body was calculated from the sum of the cumulative urinary and faecal radioactivity. The biological half decay time of albumin injected on a fat intake of 70 g per day (●------●) was two and a half days. The half decay time of albumin, two weeks (+——+), and seven weeks (O——O) after starting a low fat diet (less than 5 g per day) was eight and a half days. The half decay times were derived from the linear component of the curves. (Jeffries *et al.*, 1964.)

is likely that long-term treatment with a medium-chain triglyceride diet benefits patients with intestinal lymphangi-ectasia mainly by providing a readily assimilated calorie source (Tift & Lloyd, 1975).

Surgical treatment of intestinal lymphangiectasia is difficult in view of the diffuse nature of the lymphatic changes, but Mistilis & Skyring (1966) have found that the creation of a lymphatico-venous anastomosis using a dilated intestinal lymph vessel produced a considerable amelioration in the condition in one patient studied. The bypass reduced the

">

oedema and steatorrhoea, though the protein loss into the gastro-intestinal tract persisted. The decompression of the dilated lymph vessels resulted in a regression of the telangiectatic appearance of small intestinal mucosal lymphatics suggesting that an anatomical or functional obstruction, which was responsible for the dilation of the terminal vessels, had been overcome. It appears that surgical treatment has its best application to cases where a localised obstruction of a major intestinal lymph trunk is responsible for the syndrome.

Protein-losing enteropathy in certain forms of heart disease

Mention has already been made of the intestinal lymphangiectasia which occurs secondary to obstruction associated with various intra-abdominal pathological processes. The frequent occurrence of hypoproteinaemia and oedema in constrictive pericarditis (Davidson, Waldmann, Goodman & Gordon, 1961; Kaihara *et al.*, 1963 and Plauth *et al.*, 1964) deserves special attention.

This hypoproteinaemia and oedema have, at various times, been ascribed to frequent tapping of protein-rich ascitic fluid, malnutrition, venous congestion of the liver with impaired protein synthesis and a dilutional effect consequent on increased retention of salt and water. However, studies such as those of Plauth *et al.* (1964) using [131]I- and ^{51}Cr-labelled albumin have clearly demonstrated excessive loss of protein into the gastro-intestinal tract in constrictive pericarditis. A lymphocytopenia, indicative of lymph loss, is a frequent observation in these patients.

In addition to constrictive pericarditis, cardiac lesions which have, on occasion, been shown to be associated with protein-losing gastroenteropathy include familial cardiomyopathy (Dölle, Martini & Peterson, 1962), inter-atrial septal defect (Davidson *et al.*, 1961), Lutembacher's Syndrome, combined atrial septal defect and mitral stenosis (Stalder & Joliat, 1973) and congenital pulmonary stenosis (Jeejeebhoy, 1962).

Petersen & Hastrup (1963) have observed in a case of long-standing constrictive pericarditis that the mucosal folds and the intestinal villi were enlarged and swollen due to dilatation of the lymphatic vessels, and in the mesentery the vessel walls were slightly thickened. It is interesting that no lymphangi-ectases were observed in the stomach or large intestine. These authors demonstrated the presence of a wide spectrum of plasma proteins in the lumen of the upper small intestine.

The mechanism which leads to a protein-losing state in the intestine as a result of heart disease is debated. Two principal processes have been suggested. The inflow stasis with venous hypertension might impede the discharge of lymph into the bloodstream at the jugulo-subclavian tap. Back-pressure in the lymphatic system with incompetence of valves would result in dilatation of peripheral lymph vessels with consequent fluid and protein leakage into the tissues. Perhaps intestinal lym-phatic vessels are specially vulnerable due to lack of support from surrounding tissue. Alternatively, the venous hyperten-sion of cardiac failure might lead to splanchnic venous hyper-tension, increased capillary hydrostatic pressure and filtration and overload on the lymphatic system. Salt and water retention would aggravate the process by reducing the effective oncotic pressure in the intestinal blood capillaries.

In constrictive pericarditis in man, the thoracic duct is dilated and drainage of the thoracic duct lymph reduces the venous pressure (Dumont, Clauss, Reed & Tice, 1963). Ex-perimental constrictive pericarditis in dogs has been shown to cause a rise in pressure of the lymph in the thoracic duct with dilatation of that vessel (Blalock & Burwell, 1935). Occlusion of the superior vena cava in dogs and cats causes a rise in thoracic duct pressure and chylothorax in a large proportion of animals (Blalock, Cunningham & Robinson, 1936). Such observations are compatible with an obstruction to the dis-charge of lymph into the great veins as a result of venous inflow stasis. A study by Marshall, Neyazaki & Abrams (1965) showed that ligation of the thoracic duct in dogs leads to an increased intestinal loss of ^{131}I-labelled polyvinyl pyrrolidone and this was unaccompanied by a rise in systemic venous pressure. The response to the blocking of the thoracic duct was the development of accessory lymphatico-venous communications. Despite these new channels of lymph drainage, the intestinal permeability to the macromolecules was still increased. These results were taken to indicate that impedance of discharge of lymph into the venous system in constrictive pericarditis is the important factor in the abnormal gastro-intestinal protein loss in this condition.

On the other hand, in six cases of constrictive pericarditis, Petersen & Ottosen (1964) found a greatly increased thoracic duct lymph flow. The data obtained suggested that extra-vascular protein transport was considerably increased in constrictive pericarditis. These authors considered that increased formation of lymph in the liver and intestinal tract, due to raised hydrostatic venous pressure in the splanchnic area, is responsible for lymphatic hypertension, with a consequent transudation of fluid and protein across the mucosa of the intestine. This mechanism should operate *par excellence* in portal hypertension of chronic liver disease, but there is no clear relationship between portal hypertension and protein-losing enteropathy. No significant alterations in the metabolism of albumin were observed by Levin & Jeffay (1964) in rats with well-advanced hepatic cirrhosis. Olarte *et al.* (1964) studied the permeability of the intestine to macromolecules, by measuring the amounts of ^{131}I-labelled polyvinyl pyrrolidone excreted in the faeces following its intravenous administration. Normally 0–1.5% of the dose is recovered in a ninety-six hour faecal collection (Gordon, 1959). In a group of patients with cirrhosis and portal hypertension, values of 0–1.65% were obtained by Olarte *et al.*, suggesting no excess mucosal permeability to macromolecules. On the other hand, one third of a group of patients with cirrhosis showed a raised ^{131}I-labelled polyvinyl-pyrrolidone excretion in a study by Davcev, Vanovski, Sestakov & Tadzer (1969). In two cases where the excretion was markedly elevated, peroral biopsy of the small intestinal mucosa revealed lymphangiectatic changes. These conflicting reports require explanation: the chronicity of the portal hypertension may be important in the development of lymphatic changes, while undefined compensatory changes may operate in a proportion of cases. It is interesting that thickening of the muscle of the mesenteric and intestinal lymph vessels has been described in patients with cirrhosis or right heart failure (Reza Rafii, 1972).

Another factor which may contribute to a lymphatic hypertension in constrictive pericarditis, though not in other forms of congestive heart disease, is the involvement of the thoracic duct in the inflammatory process with stenotic changes in its lumen. Such changes have been observed radiologically (Servelle *et al.*, 1966). In one such patient, the thoracic duct appeared dilated and tortuous before pericardiectomy and these radiological changes reverted to normal, except for a single stenotic area, after the operation (Takashima & Takekoshi, 1968). The presence of permanent structural changes in the thoracic duct might explain the occasional failure of pericardiectomy to halt the protein-losing enteropathy.

In summary, the genesis of protein-losing gastro-enteropathy in constrictive pericarditis and certain other cardiac diseases is probably to be ascribed to a number of mechanisms. Increased lymph formation due to hypertension in the splanchnic vascular bed leads to an increased load of fluid and protein to be transported by the lymphatics. Dilutional hypoproteinaemia resulting from salt and water retention aggravates this situation. A rise in central venous pressure may, in some cases, oppose the entry of lymph into the great veins. Longstanding intralymphatic hypertension may lead to structural alterations in the walls of lymphatic vessels, giving rise to a functional inadequacy of these vessels as regards lymph propulsion, with consequent lymphangiectatic changes in terminal lymph vessels. The thoracic duct may become involved in a posterior mediastinitis associated with pericarditis, and stenotic changes in the vessel may ensue. The fact that a protein-losing enteropathy is not an invariable accompaniment of constrictive pericarditis is possibly a reflection of the fact that lymphatic drainage is still adequate and the chronicity of the lymphatic hypertension may play a key part in the establishment of the protein-losing enteropathy.

Whipple's disease
This rare disease of uncertain aetiology deserves special mention. It is characterised by weakness, weight loss, steatorrhoea, a variety of abdominal complaints and a polyarthropathy. Histologically, the characteristic lesion is seen in the small intestinal mucosa where numerous macrophages, which stain

strongly by the periodic acid-Schiff (PAS) technique, infiltrate the lamina propria. Similar cells are found in the mesenteric lymph nodes. Large numbers of fat droplets are found in the lamina propria of the intestinal mucosa and between the lateral cell walls of the enterocytes. An unidentified bacillus is found in the small intestinal mucosa and lymph nodes and the response of these patients to a course of antibiotic, such as tetracycline, points to an infective aetiology. The common physical signs in this condition include lymphadenopathy, fever, peripheral oedema, skin pigmentation and a mild hypotension. An 'anaemia of chronic illness', hypoalbuminaemia, hypocholesterolaemia, hypokalaemia and prothrombin deficiency are frequently present. Maizel, Ruffin & Dobbins (1970) reviewing 114 patients with this condition found steatorrhoea in 93% and decreased D-xylose absorption in 78%. However, only two of sixteen patients tested showed defective vitamin B_{12} absorption, which is in keeping with observations that the most severe lesions are found in the proximal small intestine.

Since steatorrhoea is such a prominent feature of the disease, the question arises as to the mechanism of its production. The presence of fat droplets in the lamina propria suggests a lymphatic block and, in line with this, a modest dilatation of the central lacteal is frequently seen. However, the degree of dilatation is slight in comparison with that seen in intestinal lymphangiectasia. There may therefore be some element of secondary lymphangiectasia in this condition arising from obstruction to lymph flow due to pathological changes in the mesenteric lymph nodes. The lymphatic endothelium which appears to be active in chylomicron transport (see Chapter 3) may not, however, be normal in this disease. Numerous lysosomes containing the bacilli are found in these cells and a functional impairment of the transcellular transport of fat may be a factor in the causation of the steatorrhoea. Lastly, neutral lipid synthesis in the enterocyte is probably faulty. Bacilli are found in the enterocytes and amino acid transport and fatty acid esterification by these cells are thought to be impaired. In keeping with this is the observation by Maizel, *et al.* (1970) that there was a paucity of chylomicra in the endoplasmic reticulum of the enterocytes in a patient with Whipple's disease 45 min after intraduodenal infusion of corn oil.

Experimental lymphangiectasia

Microscopic demonstration of the central lymph vessel of the intestinal villus is difficult because of its tendency to collapse during histological preparation. Papp, Röhlich, Rusznyák & Törö (1962), injected serum into the muscular coat of the cat small intestine in order to produce dilatation of the vessel for morphological studies (see Chapter 2). This could be regarded as a type of experimental lymphangiectasia.

The criteria of successful production of experimental intestinal lymphangiectasia are, first, the appearance of the characteristic morphological changes of the lymphatics and secondly, demonstration of a protein-losing gastro-enteropathy. The natural development of the condition has been described in the dog; attempts have been made to create a suitable experimental model in this animal.

Bank *et al.* (1967), however, were unable to produce histological changes of intestinal lymphangiectasia in the dog either by forced fat feeding, ligation of the thoracic duct either proximally or distally, or by blocking of regional lymphatics with sclerosing agents or by surgery. On the other hand, Marshall *et al.* (1965) reported increased loss of ^{131}I-labelled polyvinylpyrrolidone into the intestine of dogs after thoracic duct ligation, despite the fact that collateral lymphatic channels and lymphatico-venous anastomoses were opening up at the time.

The production of experimental intestinal lymphangiectasia requires a thorough interruption of the lymphatic drainage of the segment involved and, for full development of the morphological changes characteristic of this condition, the obstruction and lymphatic hypertension should continue for several weeks or months.

Techniques for producing chronic lymphatic obstruction by injection of lymph vessels and nodes with sclerosing agents were developed by Drinker, Field & Homans (1934) and used to produce experimental obstruction of intestinal lymph vessels by Reichert & Mathes (1936). These techniques have been reviewed by Kalima (1971). Chemical agents used for the purpose include crystalline silica, aqueous solutions of quinine, bismuth oxychloride or sodium morrhuate. Surgical techniques are unsuitable for producing chronic lymphatic obstruction as lymphatico-venous anastomoses and lymphatic collaterals readily open and bypass ligated vessels. Lymph node excision does not appear to be any more successful in producing chronic lymphatic obstruction in the intestine, although lymphoedema is a common result of an effective excision of the regional nodes of a limb. The anatomical arrangement of the lymphatics of the alimentary tract makes complete surgical interruption very difficult but chronic lymphatic obstruction in the ileum can be produced by combined ligation of vessels and injection of lymph nodes with sclerosing agents (Kalima, 1971).

Kalima & Collan (1970) have studied the effects of ileal lymphatic obstruction in rats. Short term obstruction, that is for a few days, resulted in hyperaemia and oedema of the intestine, with dilation of the central lymphatic vessel in the villi; the lymphatic endothelium was attenuated. Minor inflammatory changes occurred in the lamina propria, with an accumulation of macrophages. In animals with lymphatic obstruction for over one month, inflammation was much more pronounced, with collections of lymphocytes, monocytes, plasma cells and macrophages in the mucosa. These changes were most pronounced in animals in which no reorganisation of the lymphatic system had occurred to allow adequate drainage of mucosal lymph. Kalima (1971) has drawn a comparison between the histological appearances so produced and Crohn's disease. In acute intestinal lymphatic obstruction in rats, a watery diarrhoea develops within a few days associated with a protein-losing enteropathy as demonstrated by ^{131}I-labelled polyvinylpyrrolidone loss from blood to intestinal lumen, as was shown in dogs by Marshall *et al.* (1965). This must represent the transudation of protein-rich oedema fluid across the intestinal epithelium. The steatorrhoea seen in experimental lymphangiectasia and in the natural disease is due to the failure of transfer of chylomicra to the circulation via the lymph and consequent reduction in the transport of lipids across the mucosal epithelium from the intestinal lumen. Accumulation of chylomicra in the lamina propria occurs. The precise fate of these chylomicra is not clear, though some phagocytosis of lipid-like material, both by lymphatic endothelium and by macrophages, occurs.

Danese, Georgalas-Penesis, Kark & Dreiling (1972) have reported success in intestinal lymphatic blockade in dogs using the generally favoured combination of ligation and intralymphatic injection of a sclerosing agent. When lymphatic blockade involved a segment of the small intestine, a merely transient oedema was noticed, but interruption of the lymphatic drainage of the entire small intestine resulted in lymphangiectasia at all levels through the bowel wall. In a subsequent study this group showed increased loss of albumin into the gut lumen following segmental lymphatic interruption, despite the absence of lymphangiectasia on histological examination (Georgalas-Penesis *et al.* 1972).

Chylous ascites

Accumulation of lymph, or lymph-like fluid, in the peritoneal cavity has been reported in a variety of pathological conditions affecting abdominal organs and retroperitoneal tissues. The characteristic of the fluid is its high triglyceride concentration, often in excess of 500 mg per 100 ml. Centrifugation separates a clear subnatant fluid from a floating layer of chylomicra. The predisposing conditions are developmental and acquired obstructions of the gastro-intestinal lymphatic system (Frank & Kern, 1968). Such obstructive lesions can lead to rupture of lymphatic vessels commonly seen in patients with congenital megalymphatics or to transudation of lymph into the peritoneal cavity (Kinmonth, 1972). In the latter type a hypoplasia of lymphatics is usually present and a protein-losing enteropathy may develop. As in intestinal lymphangiectasia, a dietary regime in which medium-chain triglyceride replaces long-chain triglyceride has been reported to alleviate the clinical condition by reducing the dietary load on mesenteric lymph vessels which are already insufficient (Weinstein, Scanlon & Hersh, 1969).

Parenteral calcium and magnesium may be required in the management of these patients (Kinmonth, 1972).

Herbertson & Wallace (1964) reported an interesting hereditary disorder in mice in which chylous ascites developed in 11–17% of newborn heterozygous ragged (Ra$^+$) mice. The animals were normal at birth but after suckling, a milky fluid gradually accumulated in the peritoneal cavity. The wall of the small intestine was pale and thickened and its submucosa was distended with fat droplets. There were no signs of lymphatic abnormality outside the abdomen, i.e. no chylothorax or lymphoedema in limbs. Mesenteric lymph vessels were dilated and tortuous, suggesting inadequate drainage, but no anatomical defect was demonstrable and the site of entry of lymph into the peritoneal cavity was not clear.

Inflammatory disease of the gastro-intestinal tract

There is good evidence that in traumatised or inflamed tissues, lymph flow increases (see Chapter 3). In dogs, experimental inflammation of the colon produces enhanced lymph flow (Nesis & Sterns, 1973) and Sheppard & Sterns (1974) have shown that, while in normal dogs 20% of a dose of labelled albumin injected into the interstitial space of the colon is transported to the circulation by the thoracic duct in the first five hours after injection, this amount increases by approximately 50% when an acute sterile inflammation has been produced by subserosal injection of a suspension of talc. By comparison with the lymphatic route only 2–4% of the labelled albumin was transported from the colon by the blood stream in the normal or inflamed tissue. It is probable that in all forms of inflammatory bowel disease protein-rich lymph flows from the gut at greatly increased rates.

REFERENCES

Aboul-Enein, A. (1973). Changes of abdominal lymphatics in ascitic patients with schistosomal hepatic fibrosis. *J. Cardiovasc. Surg.* **14**, 529–34.

Ackermann, N. B. (1975). The influences of changes in temperature on intestinal lymph flow and relationship to operations for carcinoma of the intestine. *Surg. Gynec. & Obst.* **140**, 885–8.

Adams, E. P. (1964). 'Transport and metabolism of long chain fatty acids in the sheep.' Ph.D. Thesis. Australian National University, Canberra.

Adesola, A. O. & Schwartz, S. I. (1965). Effects of histamine on hepatic lymph production. *Surg.* **58**, 1006–12.

Aiello, R. G., Enquist, I. F., Ikezono, E. & Levowitz, B. S. (1960). An experimental study of the role of hepatic lymph in the production of ascites. *Surg. Gynec. Obstet.* **111**, 77–81.

Alexander, H. L., Shirley, K. & Allen, D. (1936). The route of ingested egg white to the systemic circulation. *J. clin. Invest.* **15**, 163–7.

Alper, C. A., Peters, J. H., Birtch, A. G. & Gardner, F. H. (1965). Haptoglobin synthesis. 1. In vivo studies of the production of haptoglobin, fibrinogen and γ-globulin by the canine liver. *J. clin. Invest.* **44**, 574–81.

Andersson, K.-E., Bergdahl, B., Dencker, H. & Wettrell, G. (1977). Activities of proscillaridin A in thoracic duct lymph after single doses in man. *Acta pharmacol. toxicol.* **40**, 280–284.

Andrews, W. H. H., Ritchie, H. D. & Maegraith, B. G. (1973). An assessment of the physiological importance of the large hepatic venous sluices in the dog. *Q. J. exp. Physiol.* **58**, 325–33.

Annegers, J. H. (1959). Absorption of lipid following thoracic duct ligation in the rat. *Am. J. Physiol.* **196**, 363–4.

Arnesjö, B., Nilsson, Å., Barrowman, J. & Borgström, B. (1969). Intestinal digestion and absorption of cholesterol and lecithin in the human. *Scand. J. Gastroenterol.* **4**, 653–65.

Asellius, G. (1627). *De Lactibus sive lacteis venis Quarto Vasorum Mesaraicorum genere novo invento.* Apud Jo Baptistam Bidellium, Mediolani.

Aune, S. (1968). The lymphatics of the stomach. In *The Physiology of Gastric Secretion* ed. L. S. Semb & J. Myren. pp. 15–17. Williams & Wilkins, Baltimore.

Azuma, T., Ohhashi, T. & Sakaguchi, M. (1977). Electrical activity of lymphatic smooth muscles. *Proc. Soc. exp. Biol. Med.* **155**, 270–3.

Azzali, G. & Didio, L. J. A. (1965). The lymphatic system of *Bradypus tridactylus. Anat. Rec.* **153**, 149–60.

Baggenstoss, A. H. & Cain, J. C. (1957). The hepatic hilar lymphatics of man. Their relation to ascites. *New Eng. J. Med.* **256**, 531–5.

Balfour, W. E. & Comline, R. S. (1962). Acceleration of the absorption of unchanged globulin in the newborn calf by factors in colostrum. *J. Physiol.* **160**, 234–57.

Bank, S., Fisher, G., Marks, I. N. & Groll, A. (1967). The lymphatics of the intestinal mucosa. A clinical and experimental study. *Am. J. dig. Dis.*, NS **12**, 619–32.

Barankay, T., Horpácsy, G., Nagy, S. & Petri, G. (1969). Changes in the level of lysosomal enzymes in plasma and lymph in hemorrhagic shock. *Med. Exp.*, *Basle* **19**, 267-71.

Baraona, E. & Lieber, C. S. (1975). Intestinal lymph formation and fat absorption: stimulation by acute ethanol administration and inhibition by chronic ethanol feeding. *Gastroenterol.* **68**, 495-502.

Barnes, J. M. & Trueta, J. (1941). Absorption of bacteria, toxins and snake venoms from the tissues: importance of the lymphatic circulation. *Lancet* **i**, 623-6.

Barrowman, J. A. & Roberts, K. B. (1967). The role of the lymphatic system in the absorption of water from the intestine of the rat. *Q. J. exp. Physiol.* **52**, 19-30.

Barrowman, J. A. & Turner, S. G. (1976). Flow and protein content of intestinal lymph during fat absorption. *J. Physiol., Lond.* **256**, 87 P-88 P.

Bartels, P. (1909). *Das Lymphgefäss System*. Gustav Fischer, Jena.

Bartos, V., Brzek, V., Groh, J. & Keller, O. (1966). Alterations in human thoracic duct lymph in relation to the function of the pancreas. *Am. J. Med. Sci.* **252**, 31-8.

Battezzati, M. & Donini, I. (1972). The internal drainage of the thoracic duct in the surgical treatment of portal hypertension. In *The Lymphatic System*, pp. 433-40. Piccin Medical Books, Padua.

Beerman, B. & Hellström, K. (1971). The efficacy of lymph drainage in the elimination of orally administered digitoxin and digoxin. *Pharmacol.* **6**, 17-21.

Bell, N. H. (1966). Comparison of intestinal absorption and esterification of 4-^{14}C vitamin D_3 and 4-^{14}C-cholesterol in the rat. *Proc. Soc. exp. Biol. Med.* **123**, 529-32.

Bell, N. H. & Bryan, P. (1969). Absorption of vitamin D_3 oleate in the rat. *Am. J. clin. Nutr.* **22**, 425-30.

Bennett, S., Shepherd, P. & Simmonds, W. J. (1962). The effect of alterations in intestinal motility induced by morphine and atropine on fat absorption in the rat. *Aust. J. exp. Biol. med. Sci.* **40**, 225-31.

Benson, J. A., Lee, P. R., Scholer, J. F., Kim, K. S. & Bollman, J. L. (1956). Water absorption from the intestine via portal and lymphatic pathways. *Am. J. Physiol.* **184**, 441-4.

Bergan, A., Taksdal, S., & Enge, I. (1975). The role of the liver lymphatics in the transport of non-erythroid bilirubin during different stages of experimental cholestasis. *Scand. J. Gastroenterol.* **10**, 571-6.

Bergan, A., Taksdal, S. & Sander, J. (1975). Transport and conjugation of ^{14}C-bilirubin during acute and chronic cholestasis in the cholecystectomized dog. *Europ. Surg. Res.* **7**, 355-65.

Bergström, S., Blomstrand, R. & Borgström, B. (1954). Route of absorption and distribution of oleic acid and triolein in the rat. *Biochem. J.* **58**, 600-4.

Bergström, K. & Werner, B. (1966). Proteins in human thoracic duct lymph. Studies on the distribution of some proteins between lymph and blood. *Acta chir. Scand.* **131**, 413-22.

Bernard, C. (1856). Mémoire sur le pancréas et sur le rôle du suc pancréatique dans les phénomènes digestifs, particulièrement dans la digestion des matières grasses neutres. *Compt. Rend. Soc. de Biol.* **43**, 379-563.

Berson, S. A. & Yalow, R. S. (1966). Insulin in blood and insulin antibodies. *Am. J. Med.* **40**, 676-90.

Biber, B., Fara, J. & Lundgren, O. (1973a). Vascular reactions in the small intestine during vasodilatation. *Acta Physiol. Scand.* **89**, 449-56.

Biber, B., Lundgren, O. & Svanvik, J. (1973). Intramural blood flow and blood volume in the small intestine of the cat as analyzed by an indicator-dilution technique. *Acta physiol. Scand.* **87**, 391-403.

Bierman, E. L., Hayes, T. L., Hawkins, J. N., Ewing, A. M. & Lindgren, F. T. (1966). Particle-size distribution of very low density plasma lipoproteins during fat absorption in man. *J. Lipid Res.* **7**, 65-72.

References

Bierman, H. R., Byron, R. L., Kelly, K. H., Gilfillan, R. S., White, L. P., Freeman, N. E. & Petrakis, N. L. (1953). The characteristics of thoracic duct lymph in man. *J. clin. Invest.* **32**, 637–49.

Biggs, M. W., Friedman, M. & Byers, S. O. (1951). Intestinal lymphatic transport of absorbed cholesterol. *Proc. Soc. exp. Biol. Med.* **78**, 641–3.

Birke, G., Liljedahl, S.-O., Plantin, L.-O. & Wetterfors, J. (1959). Ventrikelns roll in albuminkatabolism. *Nord. Med.* **62**, 1741–4.

Blalock, A. & Burwell, C. S. (1935). Thoracic duct lymph pressure in concretio cordis: experimental study. *J. lab. clin. Med.* **21**, 296–7.

Blalock, A., Cunningham, R. S. & Robinson, C. S. (1936). Experimental production of chylothorax by occlusion of the superior vena cava. *Ann. Surg.* **104**, 359–64.

Blalock, A., Robinson, C. S., Cunningham, R. S. & Gray, M. E. (1937). Experimental studies on lymphatic blockage. *Arch. Surg., Chicago* **34**, 1049–71.

Blomstrand, R. (1954). The intestinal absorption of linoleic-1-^{14}C acid. *Acta Physiol. Scand.* **32**, 99–105.

Blomstrand, R. (1955). Transport form of decanoic acid-1-^{14}C in the lymph during intestinal absorption in the rat. *Acta Physiol. Scand.* **34**, 67–70.

Blomstrand, R. & Ahrens, E. H. (1958). The absorption of fats studied in a patient with chyluria. II. Palmitic and oleic acids. *J. biol. Chem.* **233**, 321–6.

Blomstrand, R. & Forsgren, L. (1967). Intestinal absorption and esterification of vitamin D$_3$-1,2-^{3}H in man. *Acta Chem. Scand.* **21**, 1662–3.

Blomstrand, R. & Forsgren, L. (1968a). Labelled tocopherols in man. Intestinal absorption and thoracic duct lymph transport of DL-tocopheryl-3, 4 ^{14}C$_2$ acetate, DL-α-tocopheramine-3, 4,-^{14}C$_2$, DL-α-tocopherol-(5-methyl-^{3}H) and N-(methyl ^{3}H) DL-γ-tocopheramine. *Int. Z. Vitaminforsch.* **38**, 328–44.

Blomstrand, R. & Forsgren, L. (1968b). Vitamin K$_1$-^{3}H in man. Its intestinal absorption and transport in the thoracic duct lymph. *Int. Z. Vitaminforsch.* **38**, 45–64.

Blomstrand, R. & Gürtler, J. (1971). Studies on the intestinal absorption and metabolism of phytylubiquinone-(1′, 2′ -^{3}H) (hexahydroubiquinone-4) in man. *Int. J. Vit. nutr. Res.* **41**, 189–203.

Blomstrand, R. & Werner, B. (1965). Alkaline phosphatase activity in human thoracic duct lymph. *Acta chir. Scand.* **129**, 177–91.

Blomstrand, R. & Werner, B. (1967). Studies on the intestinal absorption of radioactive β-carotene and vitamin A in man. Conversion of β-carotene into vitamin A. *Scand. J. clin. Lab. Invest.* **19**, 339–45.

Blomstrand, R., Dahlbäck, O. & Radner, S. (1960). Observations on the thoracic duct lymph in patients with cirrhosis of the liver. *Acta Hepatosplen.* **7**, 1–7.

Bloom, B., Chaikoff, I. L., Reinhardt, W. O., Entenman, C. & Dauben, W. G. (1950). The quantitative significance of the lymphatic pathway in transport of absorbed fatty acids. *J. biol. Chem.* **184**, 1–8.

Bloom, B., Chaikoff, I. L. & Reinhardt, W. O. (1951). Intestinal lymph as pathway for transport of absorbed fatty acids of different chain lengths. *Am. J. Physiol.* **166**, 451–5.

Bloom, B., Kiyasu, J. Y., Reinhardt, W. O. & Chaikoff, I. L. (1954). Absorption of phospholipides–manner of transport from intestinal lumen to lacteals. *Am. J. Physiol.* **177**, 84–6.

Bloom, W. (1923). The role of lymphatics in the absorption of bile pigment from the liver in early obstructive jaundice. *Bull. Johns Hopkins Hosp.* **34**, 316–20.

Bluestein, M., Malagelada, J., Linscheer, W. & Fishman, W. H. (1970). Enzymorphology of alkaline phosphatase of rat intestine following fatty acid infusion. *J. Histochem. Cytochem.* **18**, 679.

Boak, J. L. & Woodruff, M. F. A. (1965). A modified technique for collecting mouse thoracic duct lymph. *Nature, Lond.* **205**, 396–7.

Bocklage, B. C., Doisy, E. A. Jr., Elliott, W. H. & Doisy, E. A. (1955). Absorption and metabolism of cortisone -4-^{14}C acetate. *J. biol. Chem.* **212**, 935–9.

Bocklage, B. C., Nicholas, H. J., Doisy, E. A. Jr., Elliott, W. H., Thayer, S. A. & Doisy, E. A. (1953). Synthesis and biological studies of 17-methyl-^{14}C-estradiol. *J. biol. Chem.* **202**, 27–37.

Bollman, J. L. (1948). A cage which limits the activity of rats. *J. Lab. clin. Med.* **33**, 1348.

Bollman, J. L. & Flock, E. V. (1951). Cholesterol in intestinal and hepatic lymph in the rat. *Am. J. Physiol.* **164**, 480–5.

Bollman, J. L., Cain, J. C. & Grindlay, J. H. (1948). Techniques for collection of lymph from liver, small intestine or thoracic duct of the rat. *J. Lab. clin. Med.* **33**, 1349–52.

Bolton, C. & Barnard, W. G. (1931). The pathological occurrences in liver in experimental venous stagnation. *J. Path. Bact.* **34**, 701–9.

Borgström, B. (1953). On the mechanism of intestinal fat absorption. 5. The effect of bile diversion on fat absorption in the rat. *Acta Physiol. Scand.* **28**, 279–86.

Borgström, B. (1955). Transport form of ^{14}C-decanoic acid in portal and inferior vena cava blood during absorption in the rat. *Acta Physiol. Scand.* **34**, 71–4.

Borgström, B. (1957). Studies on the phospholipids of human bile and small intestinal content. *Acta Chem. Scand.* **11**, 749.

Borgström, B. (1968*a*). A note on the intestinal absorption of cholesteryl ethers in the rat. *Proc. Soc. exp. Biol. Med.* **127**, 1120–22.

Borgström, B. (1968*b*). Quantitative aspects of the intestinal absorption and metabolism of cholesterol and beta-sitosterol in the rat. *J. Lipid Res.* **9**, 473–81.

Borgström, B. (1975). On the interactions between pancreatic lipase and colipase and the substrate and the importance of bile salts. *J. Lipid Res.* **16**, 411–17.

Borgström, B. & Laurell, C.-B. (1953). Studies on lymph and lymph-proteins during absorption of fat and saline by rats. *Acta Physiol. Scand.* **29**, 264–80.

Borgström, B. & Tryding, N. (1956). Free fatty acid content of rat thoracic duct lymph during fat absorption. *Acta Physiol. Scand.* **37**, 127–33.

Borgström, B., Dahlqvist, A., Lundh, G. & Sjövall, J. (1957). Studies of intestinal digestion and absorption in the human. *J. clin. Invest.* **36**, 1521–36.

Borgström, B., Radner, S. & Werner, B. (1970). Lymphatic transport of cholesterol in the human being. Effect of dietary cholesterol. *Scand. J. clin. Lab. Invest.* **26**, 227–35.

Boucrot, P. & Clement, J. R. (1971). Resistance to the effect of phospholipase A$_2$ of the biliary phospholipids during incubation of bile. *Lipids* **6**, 652–6.

Bouquillon, M., Carlier, H. & Clement, J. (1974). Effect of various dietary fats on the size and distribution of lymph fat particles in rat. *Digestion* **10**, 255–66.

Bowers, W. F., McKinnon, W. M., Marino, J. M. & Culverwell, J. T. (1964). Cannulation of the thoracic duct: its role in the pre-shunt management of hemorrhage due to esophageal varices. *J. Int. Coll. Surg.* **42**, 71–6.

Boyer, J. L., Maddrey, W. C., Basu, A. K. & Iber, F. L. (1969). The source of albumin in the thoracic duct lymph in patients with portal hypertension and ascites. *Am. J. med. Sci.* **257**, 32–43.

Bradham, R. R. & Takaro, T. (1968). The nature and significance of the thoracic duct-subclavian vein junction. *Surg.* **64**, 643–6.

Brambell, F. W. R. (1958). The passive immunity of the young mammal. *Biol. Rev.* **33**, 488–531.

Brambell, F. W. R. (1970). *The Transmission of Passive Immunity from Mother to Young.* North Holland Publishing Co. Amsterdam.

Brandt, J. L., Castleman, L., Ruskin, H. D., Greenwald, J. & Kelly, J. J. (1955). The effect of oral protein and glucose feeding on splanchnic blood flow and oxygen utilization in normal and cirrhotic subjects. *J. clin. Invest.* **34**, 1017–25.

References

Brauer, R. W. (1963). Liver circulation and function. *Physiol. Rev.* **43**, 115–213.

Brauer, R. W., Holloway, R. J. & Leong, G. F. (1959). Changes in liver function and structure due to experimental passive congestion under controlled hepatic vein pressures. *Am. J. Physiol.* **197**, 681–92.

Brindley, D. N. & Hübscher, G. (1966). The effect of chain length on the activation and subsequent incorporation of fatty acids into glycerides by the small intestinal mucosa. *Biochim. Biophys. Acta* **125**, 92–105.

Brown, C. S. & Hardenbergh, E. (1951). A technique for sampling lymph in unanaesthetized dogs by means of an exteriorized thoracic duct–venous shunt. *Surg.* **29**, 502–7.

Browse, N. L., Lord, R. S. A. & Taylor, A. (1971). Pressure waves and gradients in the canine thoracic duct. *J. Physiol., Lond.* **213**, 507–524.

Browse, N. L., Rutt, D. R., Sizeland, D. & Taylor, A. (1974). The velocity of lymph flow in the canine thoracic duct. *J. Physiol., Lond.* **237**, 401–13.

Bruggeman, T. M. (1975). Plasma proteins in canine gastric lymph. *Gastroenterol.* **68**, 1204–10.

Bruns, R. R. & Palade, G. E. (1968). Studies on blood capillaries. II. Transport of ferritin molecules across the wall of muscle capillaries. *J. Cell Biol.* **37**, 277–99.

Brzek, V. & Bartos, V. (1969). Therapeutic effect of the prolonged thoracic duct lymph fistula in patients with acute pancreatitis. *Digestion* **2**, 43–50.

Bullen, J. J. & Batty, I. (1957). Experimental enterotoxaemia of sheep; the effect on the permeability of the intestine and the stimulation of antitoxin production in immune animals. *J. Path. Bact.* **73**, 511–18.

Burgener, F. A. & Adams, J. T. (1976). Analysis of the radiographically visualised deep hepatic lymph drainage in the dog. *Lymphol.* **9**, 105–11.

Burkel, W. E. & Low, F. N. (1966). The fine structure of rat liver sinusoids, space of Disse and associated tissue space. *Am. J. Anat.* **118**, 769–83.

Burton, P., Hammond, E. M., Harper, A. A., Howat, H. T., Scott, J. E. & Varley, H. (1960). Serum amylase and serum lipase levels in man after administration of secretin and pancreozymin. *Gut* **1**, 125–39.

Cain, J. C., Grindlay, J. H., Bollman, J. L., Flock, E. V. & Mann, F. C. (1947). Lymph from liver and thoracic duct: An experimental study. *Surg. Gynec. Obstet.* **85**, 559–62.

Calnan, J. S., Rivero, O. R., Fillmore, S. & Mercurius-Taylor, L. (1967). Permeability of normal lymphatics. *Brit. J. Surg.* **54**, 278–85.

Campbell, R. S., Brobst, D. & Bisgard, G. (1968). Intestinal lymphangiectasia in a dog. *J. Am. vet. Med. Ass.* **153**, 1051–4.

Campbell, T. & Heath, T. (1973). Intrinsic contractility of lymphatics in sheep and in dogs. *Q. J. exp. Physiol.* **58**, 207–17.

Cardell, R. R., Badenhausen, S. & Porter, K. R. (1967). Intestinal triglyceride absorption in the rat. An electron microscopical study. *J. Cell Biol.* **34**, 123–55.

Carleton, H. M. & Florey, H. W. (1927). The mammalian lacteal: its histological structure in relation to its physiological properties. *Proc. R. Soc.* **102B**, 110–18.

Carlson, A. J. & Luckhardt, A. B. (1908). On the diastases in the blood and the body fluids. *Am. J. Physiol.* **23**, 148–64.

Carlsten, A. (1950). On the sources of the histaminase present in thoracic duct lymph. *Acta Physiol. Scand.* **20 Suppl. 70**, 5–26.

Carlsten, A., Edlund, Y. & Thulesius, O. (1961). Bilirubin, alkaline phosphatase and transaminases in blood and lymph during biliary obstruction in the cat. *Acta Physiol. Scand.* **53**, 58–67.

Carlsten, A. & Olin, T. (1952). The route of the intestinal lymph to the bloodstream. *Acta Physiol. Scand.* **25**, 259–66.

Carrier, B. (1926). Living cells in bat's wing. *Physiol. Papers (Krogh)* 1–9.

Casley-Smith, J. R. (1962). The identification of chylomicra and lipoproteins in tissue sections and their passage into jejunal lacteals. *J. Cell Biol.* **15**, 259–77.

Casley-Smith, J. R. (1964a). An electron microscopic study of injured and abnormally permeable lymphatics. *Ann. N.Y. Acad. Sci.* **116**, 803–30.

Casley-Smith, J. R. (1964b). Endothelial permeability. The passage of particles into and out of diaphragmatic lymphatics. *Q. J. exp. Physiol.* **49**, 365–83.

Casley-Smith, J. R. (1967a). The fine structure, properties and permeabilities of the lymphatic endothelium. In *New Trends in Basic Lymphology* ed. J. M. Collette, G. Jantet & E. Schoffeniels, pp. 19–39. Birkhauser Verlag, Basle & Stuttgart.

Casley-Smith, J. R. (1967b). An electron microscopical study of the passage of ions through the endothelium of lymphatic and blood capillaries and through the mesothelium. *Q. J. exp. Physiol.* **52**, 105–13.

Casley-Smith, J. R. (1969). The structure of normal large lymphatics: how this determines their permeabilities and their ability to transport lymph. *Lymphol.* **2**, 15–25.

Casley-Smith, J. R. (1971). Endothelial fenestrae in intestinal villi: differences between the arterial and venous ends of the capillaries. *Microvasc. Res.* **3**, 49–68.

Casley-Smith, J. R. (1972). The role of the endothelial intercellular junctions in the functioning of the initial lymphatics. *Angiologica* **9**, 106–31.

Casley-Smith, J. R. (1976). The functioning and interrelationships of blood capillaries and lymphatics. *Experientia* **32**, 1–12.

Casley-Smith, J. R. & Florey, H. W. (1961). The structure of normal small lymphatics. *Q. J. exp. Physiol.* **46**, 101–6.

Casley-Smith, J. R., O'Donoghue, P. J. & Crocker, K. W. J. (1975). The quantitative relationships between fenestrae in jejunal capillaries and connective tissue channels. Proof of 'tunnel-capillaries'. *Microvas. Res.* **9**, 78–100.

Chaikoff, I. L., Bloom, B., Stevens, B. P., Reinhardt, W. O. & Dauben, W. G. (1951). Pentadecanoic acid -5-^{14}C: its absorption and lymphatic transport. *J. biol. Chem.* **190**, 431–5.

Chaikoff, I. L., Bloom, B., Stevens, B. P., Reinhardt, W. O., Dauben, W. G. & Eastham, J. F. (1952). ^{14}C-cholesterol. I. Lymphatic transport of absorbed cholesterol-4-C^{14}. *J. biol. Chem.* **194**, 407–12.

Citrin, Y., Sterling, K. & Halsted, J. A. (1957). The mechanism of hypoproteinaemia associated with giant hypertrophy of the gastric mucosa. *New Eng. J. Med.* **257**, 906–12.

Clain, D. & McNulty, J. (1968). A radiological study of the lymphatics of the liver. *Br. J. Radiol.* **41**, 662–8.

Clark, E. R. (1936). Growth and development of function in blood vessels and lymphatics. *Ann. int. Med.* **9**, 1043–9.

Clark, E. R. & Clark, E. L. (1937). Observations on living mammalian lymphatic capillaries – their relation to the blood vessels. *Am. J. Anat.* **60**, 253–298.

Clark, S. B. & Norum, K. R. (1977). The lecithin cholesterol acyl transferase activity of rat intestinal lymph. *J. Lip. Res.* **18**, 293–300.

Clementi, F. & Palade, G. E. (1969). Intestinal capillaries. 1. Permeability to peroxidase and ferritin. *J. Cell Biol.* **41**, 33–58.

Clendinnen, B. G., Reeder, D. D. & Thompson, J. C. (1973). The role of thoracic duct lymph in gastrin transport and gastric secretion. *Gut* **14**, 30–4.

Coates, M. E., Thompson, S. Y. & Kon, S. K. (1950). Conversion of carotene to vitamin A in the intestine of the pig and of the rat: transport of vitamin A by the lymph. *Biochem. J.* **46**, 30 P–31 P.

Code, C. F., Bass, P., McClary, G. B., Newnum, R. L. & Orvis, A. L. (1960). Absorption of water, sodium and potassium in small intestine of dogs. *Am. J. Physiol.* **199**, 281–8.

References

Code, C. F. & Pickard, D. W. (1973). The importance of the lymphatic system in the absorption of water from the intestine. *J. Physiol:, Lond.* **231**, 40P–41P.

Cole, W. R., Witte, M. H., Kash, S. L., Rodger, M., Bleisch, V. R. & Muelheims, G. H. (1967). Thoracic duct to pulmonary vein shunt in the treatment of experimental right heart failure. *Circulation* **36**, 539–43.

Cole, W. R., Petit, R., Brown, A. & Witte, M. H. (1968). Lymphatic transport of bacteria in surgical infection. *Lymphol.* **1**, 52–7.

Collan, Y. & Kalima, T. V. (1970). The lymphatic pump of the intestinal villus of the rat. *Scand. J. Gastroenter.* **5**, 187–96.

Collan, Y. & Kalima, T. V. (1974). Topographic relations of lymphatic endothelial cells in the initial lymphatic of the intestinal villus. *Lymphol.* **7**, 175–84.

Comline, R. S., Roberts, H. E. & Titchen, D. A. (1951a). Route of absorption of colostrum globulin in the new-born animal. *Nature, Lond.* **167**, 561–2.

Comline, R. S., Roberts, H. E. & Titchen, D. A. (1951b). Histological changes in the epithelium of the small intestine during protein absorption in the newborn animal. *Nature, Lond.* **168**, 84–5.

Comparini, L. (1969). Lymph vessels of the liver in man. Microscopic morphology and histotopography. *Angiologica* **6**, 262–74.

Co Tui, F., Barcham, I. S. & Shafiroff, B. G. P. (1944). Ligation of the thoracic duct and the posthaemorrhage plasma protein level. *Surg. Gynec. Obstet.* **79**, 37–40.

Courtice, F. C., & Morris, B. (1955). The exchange of lipids between plasma and lymph of animals. *Q. J. exp. Physiol.* **40**, 138–48.

Courtice, F. C., Simmonds, W. J. & Steinbeck, A. W. (1951). Some investigations on lymph from a thoracic duct fistula in man. *Aust. J. exp. Biol. Med. Sci.* **29**, 201–10.

Courtice, F. C., Woolley, G. & Garlick, D. G. (1962). The transference of macromolecules from plasma to lymph in the liver. *Aust. J. exp. Biol.* **40**, 111–19.

Crandall, L. A., Barker, S. B. & Graham, D. G. (1943). Study of the lymph flow from a patient with thoracic duct fistula. *Gastroenterol.* **1**, 1040–8.

Cruickshank, W. (1786). *The Anatomy of the Absorbing Vessels of the Human Body.* G. Nicol, London.

Cueto, J. & Currie, R. A. (1967). Cannulation of the thoracic duct and umbilical vein in patients with portal hypertension. *Ann. Surg.* **165**, 408–14.

Dahlbäck, O., Hansson, R., Tibbling, G. & Tryding, N. (1968). The effect of heparin on diamine oxidase and lipoprotein lipase in human lymph and blood plasma. *Scand. J. clin. Lab. Invest.* **21**, 17–25.

Danese, C. A., Georgalas-Penesis, M., Kark, A. E. & Dreiling, D. A. (1972). Studies of the effects of blockage of intestinal lymphatics. I. Experimental procedure and structural alterations. *Am. J. Gastroenterol.* **57**, 541–6.

Daniel, P. M. & Henderson, J. R. (1966). Insulin in the lymph of the thoracic duct of the rabbit. *J. Physiol., Lond.* **184**, 36P–37P.

Daniel, P. M., Gale, M. M. & Pratt, O. E. (1963). Radioactive iodine in the lymph leaving the thyroid gland. *Q. J. exp. Physiol.* **48**, 138–45.

Datta, D. V. (1969). In *Monogram on Non-cirrhotic portal fibrosis*. Indian Council of Medical Research, New Delhi.

Datta, D. V., Samanta, A. K. S., Patra, B. S., Saini, V. K. & Chhuttani, P. N. (1971). Management of bleeding oesophageal varices by draining lymph from the thoracic duct. *Gut* **12**, 48–50.

Davcev, P., Vanovski, B., Sestakov, D. & Tadzer, I. (1969). Protein-losing enteropathy in patients with liver cirrhosis. *Dig.* **2**, 17–22.

Davidson, J. D., Waldmann, T. A., Goodman, D. S. & Gordon, R. S. (1961). Protein-losing gastro-enteropathy in congestive heart failure. *Lancet* **i**, 899–902.

Dawson, A. M. & Isselbacher, K. J. (1960). Studies on lipid metabolism in the small intestine with observations on the role of bile salts. *J. clin. Invest.* **39**, 730–40.

Deane, H. W. (1964). Some electron microscopic observations on the lamina propria of the gut with comments on the close association of macrophages, plasma cells and eosinophils. *Anat. Rec.* **149**, 453–74.

Dedo, D. D. & Ogura, J. H. (1975). Exteriorization of thoracic duct lymph. *Arch. Otolaryngol.* **101**, 671–4.

De Haas, G. H., Postema, N. M., Nieuwenhuizen, W. & Van Deenen, L. L. M. (1968). Purification and properties of phospholipase A from porcine pancreas. *Biochim. Biophys. Acta* **159**, 103–17.

Del Rio Lozano, I. & Andrews, W. H. H. (1966). Some relationships between the lymphatic system, the biliary system and the blood vascular systems in the liver of the rabbit. *Q. J. exp. Physiol.* **51**, 324–35.

De Marco, T. J. & Levine, R. R. (1969). Role of the lymphatics in the intestinal absorption and distribution of drugs. *J. Pharmacol. exp. Therap.* **169**, 142–51.

Denis, G. (1968). Ammonia content of canine lymph. *Rev. Canad. Biol.* **27**, 115–120.

Deysine, M., Mader, M., Rosario, E. & Mandell, C. (1974). Lymphatic flow alterations secondary to changes in total serum calcium levels. *Proc. Soc. exp. Biol. Med.* **147**, 158–61.

Dietschy, J. M. & Siperstein, M. D. (1965). Cholesterol synthesis by the gastro-intestinal tract: Localization and mechanisms of control. *J. clin. Invest.* **44**, 1311–27.

Dive, Ch. C., Nadalini, A. C., & Heremans, J. F. (1971). Origin and composition of hepatic lymph proteins in the dog. *Lymphol.* **4**, 133–9.

Dobbins, W. O. (1966a). The intestinal mucosal lymphatic in man. A light and electron microscopic study. *Gastroenterol.* **51**, 994–1003.

Dobbins, W. O. (1966b). Electron microscopic study of the intestinal mucosa in intestinal lymphangiectasia. *Gastroenterol.* **51**, 1004–17.

Dobbins, W. O. (1968). Hypo-β-lipoproteinaemia and intestinal lymphangiectasia. A new syndrome of malabsorption and protein-losing enteropathy. *Arch. int. Med.* **122**, 31–8.

Dobbins, W. O. (1971). Intestinal mucosal lacteal in transport of macromolecules and chylomicrons. *Am. J. clin. Nutr.* **24**, 77–90.

Dobbins, W. O. & Rollins, E. L. (1970). Intestinal mucosal lymphatic permeability: an electron microscopic study of endothelial vesicles and cell junctions. *J. Ultrastruc. Res.* **33**, 29–59.

Dodd, G. D. (1970). Lymphography in diseases of the liver and pancreas. *Radiol. Clin. N. Am.* **8**, 69–84.

Doemling, D. B. & Steggerda, F. R. (1960). Chronic thoracic duct–venous shunt preparation in dogs. *J. appl. Physiol.* **15**, 745–6.

Dogiel, A. (1883). Über die Beziehungen zwischen Blut- und Lymphgefässen. *Arch. Mikrosk. Anat. EntwMech. Org.* **22**, 608–15.

Dölle, W., Martini, G. A. & Petersen, F. (1962). Idiopathic familial cardiomegaly with intermittent loss of protein into the gastro-intestinal tract. *Ger. Med. Monthly* **7**, 300–6.

Donini, I. & Bresadola, F. (1974). Cervical lymph–venous shunt in experimental ascites and in patients with hepatic cirrhosis. *Lymphol.* **7**, 105–8.

Dorigotti, L. & Glässer, A. H. (1968). Comparative effects of caerulein, pancreozymin and secretin on pancreatic blood flow. *Experientia* **24**, 806–7.

Dowling, R. H. (1972). The enterohepatic circulation. *Gastroenterol.* **62**, 122–40.

Drinker, C. K. (1946). Extravascular protein and the lymphatic system. *Ann. N.Y. Acad. Sci.* **46**, 807–21.

Drinker, C. K., Field, M. E. & Homans, J. (1934). The experimental production of edema and elephantiasis as a result of lymphatic obstruction. *Am. J. Physiol.* **108**, 509–20.

References

Drinker, C. K., Field, M. E. & Ward, H. K. (1934). The filtering capacity of lymph nodes. *J. exp. Med.* **59**, 393–405.

Drummond, J. C., Bell, M. E. & Palmer, E. T. (1935). Observations on the absorption of carotene and vitamin A. *Br. Med. J.* **1**, 1208–10.

Dumont, A. E. (1973). Icteric thoracic duct lymph. Significance in patients with manifestations of obstructive jaundice. *Ann. Surg.* **178**, 53–8.

Dumont, A. E. (1975a). Liver lymph. In *The liver–normal and abnormal functions*, Part A, ed. Frederick F. Becker, Chap. 3. Dekker, New York.

Dumont, A. E. (1975b). The flow capacity of the thoracic duct–venous junction. *Am. J. med. Sci.* **269**, 292–301.

Dumont, A. E. & Martelli, A. B. (1969). X-ray opacification of hepatic lymph nodes following intravenous injection of tantalum dust. *Lymphol.* **2**, 91–5.

Dumont, A. E. & Mulholland, J. H. (1960). Flow rate and composition of thoracic-duct lymph in patients with cirrhosis. *New Engl. J. Med.* **263**, 471–4.

Dumont, A. E. & Mulholland, J. H. (1962). Alterations in thoracic duct lymph flow in hepatic cirrhosis: Significance in portal hypertension. *Ann. Surg.* **156**, 668–77.

Dumont, A. E. & Weissmann, G. (1964). Lymphatic transport of beta-glucuronidase during haemorrhagic shock. *Nature, Lond.* **201**, 1231–2.

Dumont, A. E. & Witte, M. H. (1966). Contrasting patterns of thoracic duct lymph formation in hepatic cirrhosis. *Surg Gynec. Obstet.* **122**, 524–8.

Dumont, A. E. & Witte, M. H. (1969). Clinical usefulness of thoracic duct cannulation. *Adv. intern. Med.* **15**, 51–71.

Dumont, A. E., Clauss, R. H., Reed, G. E. & Tice, D. A. (1963). Lymph drainage in patients with congestive heart failure. *New Eng. J. Med.* **269**, 949–52.

Dumont, A. E., Doubilet, H. & Mulholland, J. H. (1960). Lymphatic pathway of pancreatic secretion in man. *Ann. Surg.* **152**, 403–9.

Dumont, A. E., Doubilet, H., Witte, C. L. & Mulholland, J. H. (1961). Disorders of the biliary-pancreatic system: Observations on lymph drainage in jaundiced patients. *Ann. Surg.* **153**, 774–80.

Dumont, A. E., Witte, C. L. & Witte, M. H. (1975). Protein content of liver lymph in patients with portal hypertension secondary to hepatic cirrhosis. *Lymphol.* **8**, 111–13.

Dumont, A. E., Witte, C. L., Witte, M. H. & Cole, W. R. (1970). Origin of red blood cells in thoracic duct lymph in hepatic cirrhosis. *Ann. Surg.* **171**, 1–8.

Dupont, J. M. & Litvine, J. (1964). Le facteur lymphatique dans les pancréatites expérimentales. *Acta Chir. Belg.* **63**, 687–97.

Duprez, A., Godart, S., Platteborse, R., Litvine, J. & Dupont, J. M. (1963). La voie de dérivation interstitielle et lymphatique de la sécrétion exocrine du pancréas. *Bull. Acad. Roy. Med. Belg.* **3**, 691–706.

Eden, E. & Sellers, K. C. (1949). The absorption of vitamin A in ruminants and rats. *Biochem. J.* **44**, 264–7.

Edkins, J. S. (1906). The chemical mechanism of gastric secretion. *J. Physiol., Lond.* **34**, 133–44.

Edlund, Y., Ekholm, R. & Zelander, T. (1963). The effect of intraductal pressure increase on the ultrastructure of the pancreas. *Acta chir. Scand.* **125**, 529–30.

Egdahl, R. H. (1958). Mechanism of blood enzyme changes following the production of experimental pancreatitis. *Ann. Surg.* **148**, 389–99.

El-Gendi, M. A., & Zaky, H. A. (1970). Thoracic duct lymph flow: A theoretical concept. *Surg.* **68**, 786–90.

El-Zawahry, M. D., El-Roubi, O. A. & Ibrahim, A. S. (1977). Protein content of thoracic duct lymph in patients with bilharzial hepatic fibrosis. *Gastroenterol.* **72**, 617–20.

Elkes, J. J. & Frazer, A. C. (1943). The relationship of phospholipin to the absorption of unhydrolysed fat from the intestine. *J. Physiol., Lond.* **102**, 24P–25P.

282

Endicott, K. M., Gillman, T., Brecher, G., Ness, A. T., Clarke, F. A. & Adamik, E. R. (1949). A study of histochemical iron using tracer methods. *J. lab. clin. Med.* **34**, 414–21.

Erlanson, C. & Borgström, B. (1968). The identity of vitamin A esterase activity of rat pancreatic juice. *Biochim. Biophys. Acta* **167**, 629–31.

Erlanson, C. & Borgström, B. (1970). Carboxylester hydrolase and lipase of human pancreatic juice and intestinal content. Behaviour in gel filtration. *Scand. J. Gastroenterol.* **5**, 395–400.

Everett, N. B., Garrett, W. E. & Simmons, B. S. (1954). Lymphatics in iron absorption and transport. *Am. J. Physiol.* **178**, 45–8.

Fara, J. W. & Madden, K. S. (1975). Effect of secretin and cholecystokinin on small intestinal blood flow distribution. *Am. J. Physiol.* **229**, 1365–70.

Fara, J. W., Rubinstein, E. H. & Sonnenschein, R. R. (1969). Visceral and behavioural responses to intraduodenal fat. *Science*, **166**, 110–11.

Fara, J. W., Rubinstein, E. H. & Sonnenschein, R. R. (1972). Intestinal hormones in mesenteric vasodilation after intraduodenal agents. *Am. J. Physiol.* **223**, 1058–67.

Faulk, W. P., McCormick, J. N., Goodman, J. R., Yoffey, J. M. & Fudenberg, H. H. (1971). Peyer's patches: morphological studies. *Cell. Immunol.* **1**, 500–20.

Felinski, L., Garton, G. A., Lough, A. K. & Phillipson, A. T. (1964). Lipids of sheep lymph. Transport from the intestine. *Biochem. J.* **90**, 154–60.

Feng, T. P., Hou, H. C. & Lim, R. K. S. (1929). Mechanism of inhibition of gastric secretion by fat. *Ch. J. Physiol.* **3**, 371–80.

Fernandes, J., van de Kamer, J. H. & Weijers, H. A. (1955). The absorption of fats studied in a child with a chylothorax. *J. clin. Invest.* **34**, 1026–36.

Fidge, N. H., Shiratori, T., Ganguly, J. & Goodman, D. S. (1968). Pathways of absorption of retinal and retinoic acid in the rat. *J. Lipid Res.* **9**, 103–9.

Field, M. E., Leigh, O. C., Heim, J. W. & Drinker, C. K. (1934–5). The protein content and osmotic pressure of blood serum and lymph from various sources in the dog. *Am. J. Physiol.* **110**, 174–81.

Fish, J. C., Sarles, H. E., Mattingly, A. T., Ross, M. V. & Remmers, A. R. (1969). Preparation of chronic thoracic duct lymph fistulas in man and laboratory animals. *J. surg. Res.* **9**, 101–6.

Fisher, R. B. & Parsons, D. S. (1949). A preparation of surviving rat small intestine for the study of absorption. *J. Physiol., Lond.* **110**, 36–46.

Fleischl, E. (1874). Von der Lymphe und den Lymphgefässen der Leber. *Ber. Verh. Sächs. Akad. Wiss., Math-Phys. Kl.* **26**, 42–55.

Flock, E. V. & Bollman, J. L. (1948). Alkaline phosphatase in the intestinal lymph of the rat. *J. biol. Chem.* **175**, 439–49.

Flock, E. V. & Bollman, J. L. (1950). The influence of bile on the alkaline phosphatase activity of intestinal lymph. *J. biol. Chem.* **184**, 523–8.

Florey, H. W. (1927a). Observations on the contractility of lacteals. I. *J. Physiol., Lond.* **62**, 267–72.

Florey, H. W. (1927b). Observations on the contractility of lacteals. II. *J. Physiol., Lond.* **63**, 1–18.

Földi, M., Kepes, J. & Szabó, G. (1954). Unpublished data quoted in Rusznyak, I., Földi, M. & Szabó, G. (1967). *Lymphatics and Lymph Circulation*, 2nd edn. Pergamon Press, Oxford.

Folkow, B. (1967). Regional adjustments of intestinal blood flow. *Gastroenterol.* **52**, 423–32.

Folkow, B., Lewis, D. H., Lundgren, O., Mellander, S. & Wallentin, I. (1964). The effect of graded vasoconstrictor fibre stimulation on the intestinal resistance and capacitance vessels. *Acta physiol. Scand.* **61**, 445–57.

Follansbee, R. (1945). The osmotic activity of gastro-intestinal fluids after water ingestion in the rat. *Am. J. Physiol.* **144**, 355–62.

References

Forbes, G. B. (1944). Chylothorax in infancy; observations on absorption of vitamins A and D and on intravenous replacement of aspirated chyle. *J. Pediat.* **25**, 191–200.

Forker, L. L., Chaikoff, I. L. & Reinhardt, W. O. (1952). Circulation of plasma proteins: their transport to lymph. *J. biol. Chem.* **197**, 625–36.

Forsgren, L. (1969). Studies on the intestinal absorption of labelled fat-soluble vitamins (A, D, E and K) via the thoracic duct lymph in the absence of bile in man. *Acta chir. Scand.* **Suppl. 399**.

Frank, B. W. & Kern, F. (1968). Intestinal and liver lymph and lymphatics. *Gastroenterol.* **55**, 408–22.

Frank, B. W., Kern, F., Franks, J. J. & Urban, E. (1966). Failure of medium-chain triglycerides in the treatment of persistent chylous ascites secondary to lymphosarcoma. *Gastroenterol.* **50**, 677–83.

Fraser, R. (1970). Size and lipid composition of chylomicrons of different Svedberg units of flotation. *J. Lipid Res.* **11**, 60–5.

Fraser, R. & Courtice, F. C. (1969). The transport of cholesterol in thoracic duct lymph of animals fed cholesterol with varying triglyceride loads. *Aust. J. exp. Biol. med. Sci.* **47**, 723–32.

Fraser, D. R. & Kodicek, E. (1968). Investigations on vitamin D esters synthesized in rats. *Biochem. J.* **106**, 485–96.

Fraser, R., Cliff, W. J. & Courtice, F. C. (1968). The effect of dietary fat load on the size and composition of chylomicrons in thoracic duct lymph. *Q. J. exp. Physiol.* **53**, 390–8.

Freeman, L. W. (1942). Lymphatic pathways from the intestine in the dog. *Anat. Rec.* **82**, 543–50.

Frick, P. G., Riedler, G. & Brögli, H. (1967). Dose response and minimal daily requirement for vitamin K in man. *J. app. Physiol.* **23**, 387–9.

Friedman, H. I. & Cardell, R. R. (1972). Effects of puromycin on the structure of rat intestinal epithelial cells during fat absorption. *J. Cell Biol.* **52**, 15–40.

Friedman, M., Byers, S. O. & Omoto, C. (1956). Some characteristics of hepatic lymph in the intact rat. *Am. J. Physiol.* **184**, 11–17.

Fronek, K. & Stahlgren, L. H. (1968). Systemic and regional hemodynamic changes during food intake and digestion in non-anaesthetized dogs. *Circulation Res.* **23**, 687–92.

Fujii, J. & Wernze, H. (1966). Effect of vasopressor substances on the thoracic duct lymph flow. *Nature, Lond.* **210**, 956–7.

Gabler, W. L. & Fosdick, L. S. (1964). Flow rate and composition of thoracic duct lymph with and without cisterna chyli ligation. *Proc. Soc. exp. Biol. Med.* **115**, 915–18.

Gabrio, B. W. & Salomon, K. (1950). Distribution of total ferritin in intestine and mesenteric lymph nodes of horses after iron feeding. *Proc. Soc. exp. Biol. Med.* **75**, 124–7.

Gage, S. H. (1920). The free granules (chylomicrons) of the blood as shown by the dark-field microscope. *Cornell Vet.* **10**, 154–5.

Gagnon, M. & Dawson, A. M. (1968). The effect of bile on vitamin A absorption in the rat. *Proc. Soc. exp. Biol. Med.* **127**, 99–102.

Gallagher, N., Webb, J. & Dawson, A. M. (1965). The absorption of ^{14}C oleic acid and ^{14}C triolein in bile fistula rats. *Clin. Sci.* **29**, 73–82.

Gallo-Torres, H. E. (1970*a*). Obligatory role of bile for the intestinal absorption of vitamin E. *Lipids* **5**, 379–84.

Gallo-Torres, H. E. (1970*b*). Intestinal absorption and lymphatic transport of D, 1-3, 4-$^{3}H_2$ α-tocopheryl nicotinate in the rat. *Int. Z. Vitaminforsch* **40**, 505–14.

Gallo-Torres, H. E. & Miller, O. N. (1969). Some aspects of lymph production in the rat. *Proc. Soc. exp. Biol. Med.* **132**, 1–5.

Gans, H. & Matsumoto, K. (1974). Are enteric endotoxins able to escape from the intestine? *Proc. Soc. exp. Biol. Med.* **147**, 736–9.

References

García-Alvarez, J., Reverte-Cejudo, D., Bernaldo-de-Quirós-González, J. & Valle-Jiménez, A. (1974). Intestinal lymphangiectasia. Report of a case with asymmetrical jaundice. *Dig.* **10**, 73–81.

Garlick, D. G. & Renkin, E. M. (1970). Transport of large molecules from plasma to interstitial fluid and lymph in dogs. *Am. J. Physiol.* **219**, 1595–605.

Georgalas-Penesis, M., Danese, C. A., Wastell, C., Kark, A. E. & Dreiling, D. A. (1972). Studies of the effects of blockade of intestinal lymphatics. II. Metabolic considerations. *Am. J. Gastroenterol.* **58**, 15–21.

Gesner, B. M. & Gowans, J. L. (1962). The output of lymphocytes from the thoracic duct lymph of unanaesthetized mice. *Br. J. exp. Path* **43**, 424–30.

Giannina, T., Steinetz, B. G. & Meli, A. (1966). Pathway of absorption of orally administered ethynyl estradiol-3-cyclopentyl ether in the rat as influenced by vehicle of administration. *Proc. Soc. exp. Biol. Med.* **121**, 1175–9.

Gibbs, G. E. & Ivy, A. C. (1951). Early histological changes following obstruction of pancreatic ducts in dogs: correlation with serum amylase. *Proc. Soc. exp. Biol. Med.* **77**, 251–4.

Gillman, T. & Ivy, A. C. (1947). A histological study of the participation of the intestinal epithelium, the reticuloendothelial system and the lymphatics in iron absorption and transport. *Gastroenterol.* **9**, 162–9.

Girardet, R. E. (1975). Surgical techniques for long-term studies of thoracic duct circulation in the rat. *J. appl. Physiol.* **39**, 682–8.

Glenn, W. W. L., Cresson, S. L., Bauer, F. X., Goldstein, F., Hoffman, O. & Healey, J. E. (1949). Experimental thoracic duct fistula. *Surg. Gynec. Obstet.* **89**, 200–8.

Glickman, R. M. & Kirsch, K. (1973). Lymph chylomicron formation during the inhibition of protein synthesis. Studies of chylomicron apoproteins. *J. clin. Invest.* **52**, 2910–20.

Glickman, R. M., Alpers, D. H., Drummey, G. D. & Isselbacher, K. J. (1970). Increased lymph alkaline phosphatase after fat feeding: effects of medium chain triglycerides and inhibition of protein synthesis. *Biochim. Biophys. Acta* **201**, 226–35.

Glickman, R. M., Kirsch, K. & Isselbacher, K. J. (1972). Fat absorption during inhibition of protein synthesis: studies of lymph chylomicrons. *J. clin. Invest.* **51**, 356–63.

Glickman, R. M., Perotto, J. L. & Kirsch, K. (1976). Intestinal lipoprotein formation: effect of colchicine. *Gastroenterol.* **70**, 347–52.

Godart, S. (1965). Lymphatic circulation of the pancreas. *Bibl. Anat.* **7**, 410–13.

Gold, R. H. & Youker, J. E. (1973). Idiopathic intestinal lymphangiectasis (primary protein-losing enteropathy). Lymphographic verification of enteric and peritoneal leakage of chyle. *Radiology* **109**, 315–16.

Gonzalez-Oddone, M. V. (1946). Bilirubin, bromsulfalein, bile acids, alkaline phosphatase and cholesterol of thoracic duct lymph in experimental regurgitation jaundice. *Proc. Soc. exp. Biol. Med.* **63**, 144–7.

Goodman, DeW. S., Blomstrand, R., Werner, B., Huang, H. S. & Shiratori, T. (1966). The intestinal absorption and metabolism of vitamin A and β-carotene in man. *J. clin. Invest.* **45**, 1615–23.

Gordon, R. S. (1959). Exudative enteropathy. Abnormal permeability of the gastro-intestinal tract demonstrable with labelled polyvinylpyrrolidone. *Lancet* **i**, 325–6.

Gotto, A. M., Levy, R. I., John, K. & Fredrickson, D. S. (1971). On the protein defect in a-β-lipoproteinemia. *New England J. Med.* **284**, 813–18.

Gowans, J. L. (1957). The effect of the continuous re-infusion of lymph and lympho-cytes on the output of lymphocytes from the thoracic duct of unanaesthetized rats. *Br. J. exp. Path.* **38**, 67–78.

Granger, D. N., Mortillaro, N. A. & Taylor, A. E. (1977). Interactions of intestinal lymph flow and secretion. *Am. J. Physiol.* **232**, E13–E18.

References

Grau, H. & Taher, E. (1965). Histologische Untersuchungen über das innere LymphGefäss-System von Pancreas und Milz. *Berliner Münchener Tierarzt. Wochenschr.* **78**, 147–51.

Greenway, C. V. & Lautt, W. W. (1970). The effects of hepatic venous pressure on trans-sinusoidal fluid transfer in the liver of the anaesthetized cat. *Circ. Res.* **26**, 697–703.

Greenway, C. V. & Murthy, V. S. (1972). Effects of vasopressin and isoprenaline infusions on the distribution of blood flow in the intestine: criteria for the validity of microsphere studies. *Br. J. Pharmacol.* **46**, 177–88.

Greenway, C. V. & Stark, R. D. (1971). Hepatic vascular bed. *Physiol. Rev.* **51**, 23–65.

Gregory, H., Hardy, P. M., Jones, D. S., Kenner, G. W. & Sheppard, R. C. (1964). The antral hormone gastrin: structure of gastrin. *Nature, Lond.* **204**, 931–3.

Griffen, W. O., Belin, R. P., Furman, R. W., Lieber, A., Schaefer, J. W. & Dubilier, L. D. (1972). Colonic lymphangiectasia: report of two cases. *Dis. Co. Rect.* **15**, 49–54.

Grim, E. & Lindseth, E. O. (1958). Distribution of blood flow to the tissues of the small intestine of the dog. *Univ. Minn. Med. Bull.* **30**, 138–45.

Grotte, G. (1956). Passage of dextran molecules across the blood–lymph barrier. *Acta chir. Scand., Suppl.* **211**, 1–84.

Grotte, G., Juhlin, L. & Sandberg, N. (1960). Passage of solid spherical particles across the blood–lymph barrier. *Acta Physiol. Scand.* **50**, 287–93.

Gullberg, R. & Olhagen, B. (1959). Electrophoresis of human gastric juice. *Nature* **184**, 1848–9.

Gurr, M. I., Brindley, D. N. & Hübscher, G. (1965). Metabolism of phospholipids. VIII. Biosynthesis of phosphatidyl choline in the intestinal mucosa. *Biochim. Biophys. Acta* **98**, 486–501.

Guyton, A. C. (1963). A concept of negative interstitial pressure based on pressures in implanted perforated capsules. *Circulation Res.* **12**, 399–414.

Guyton, A. C., Granger, H. T. & Taylor, A. E. (1971). Interstitial fluid pressure. *Phys. Rev.* **51**, 527–63.

Hadorn, B., Sumida, Ch., Tarlow, M. J., Robinson, L. & White, T. T. (1971). The activation of human pancreatic proteinases. Studies on human pancreatic juice and duodenal juice of four children with intestinal enterokinase deficiency. *Acta Paed. Scand.* **60**, 369.

Haljamäe, H., Jodal, M., Lundgren, O. & Svanvik, J. (1971). The distribution of sodium in intestinal villi during absorption of sodium chloride. *Acta physiol. Scand.* **83**, 283–5.

Haljamäe, H., Jodal, M. & Lundgren, O. (1973). Countercurrent multiplication of sodium in intestinal villi during absorption of sodium chloride. *Acta physiol. Scand.* **89**, 580–93.

Hall, J. G., Morris, B. & Woolley, G. (1965). Intrinsic rhythmic propulsion of lymph in the unanaesthetized sheep. *J. Physiol., Lond.* **180**, 336–49.

Hamosh, M. & Scow, R. O. (1973). Lingual lipase and its role in the digestion of dietary lipid. *J. clin. Invest.* **52**, 88–95.

Hamosh, M., Klaeveman, H. L., Wolf, R. O. & Scow, R. (1975). Pharyngeal lipase and digestion of dietary triglyceride in man. *J. clin. Invest.* **55**, 908–13.

Hanzon, V. (1952). Liver cell secretion under normal and pathologic conditions studied by fluorescence microscopy on living rats. *Acta Physiol. Scand.* **28, suppl.** 101.

Hardy, R. N. (1969). The influence of specific chemical factors in the solvent on the absorption of macromolecular substances from the small intestine of the newborn calf. *J. Physiol., Lond.* **204**, 607–32.

Hargens, A. R. & Zweifach, B. W. (1976). Transport between blood and peripheral lymph in intestine. *Microvasc. Res.* **11**, 89–101.

References

Hartmann, P. E. & Lascelles, A. K. (1966). The flow and lipoid composition of thoracic duct lymph in the grazing cow. *J. Physiol., Lond.* **184**, 193–202.

Hatch, F. T., Aso, Y., Hagopian, L. M. & Rubenstein, J. L. (1966). Biosynthesis of lipoprotein by rat intestinal mucosa. *J. biol. Chem.* **241**, 1655–65.

Haussler, M. R., Nagode, L. A. & Rasmussen, H. (1970). Induction of intestinal brush border alkaline phosphatase by vitamin D and identity with Ca-ATPase. *Nature, Lond.* **228**, 1199–201.

Heath, T. (1964). Pathways of intestinal lymph drainage in normal sheep and in sheep following thoracic duct occlusion. *Am. J. Anat.* **115**, 569–80.

Heaton, K. W. (1972). *Bile Salts in Health and Disease.* Churchill Livingstone, Edinburgh and London.

Heidenhain, R. (1891). Versuche und Fragen zur Lehre von der Lymphbildung. *Pflügers Arch. ges. Physiol.* **49**, 209–301.

Hellman, L., Bradlow, H. L., Frazell, E. L. & Gallagher, T. F. (1956). Tracer studies of the absorption and fate of steroid hormones in man. *J. clin. Invest.* **35**, 1033–44.

Hendrix, B. M. & Sweet, J. E. (1917). A study of amino nitrogen and glucose in lymph and blood before and after the injection of nutrient solutions in the intestine. *J. biol. Chem.* **32**, 299–307.

Henry, C. G. (1933). Studies on the lymphatic vessels and on the movement of lymph in the ear of the rabbit. *Anat. Rec.* **57**, 263–78.

Herbertson, B. M. & Wallace, M. E. (1964). Chylous ascites in newborn mice. *J. med. Genet.* **1**, 10–23.

Hernell, O. & Olivecrona, T. (1974). Human milk lipases. 1. Serum-stimulated lipase. *J. Lipid Res.* **15**, 367–74.

Herrick, J. F., Essex, H. E., Mann, F. C. & Baldes, E. J. (1934). The effect of digestion on blood flow in certain blood vessels of the dog. *Am. J. Physiol.* **108**, 621–8.

Herring, P. T. & Simpson, S. (1909). The pressure of pancreatic secretion and the mode of absorption of pancreatic juice after obstruction of the main ducts of the pancreas. *Q. J. exp. Physiol.* **2**, 99–108.

Hewson, W. (1774). A description of the lymphatic system in the human subject and in other animals. In *The Works of William Hewson*, ed. G. Gulliver. Sydenham Society, London.

Hiatt, N. & Bonorris, G. (1966). Removal of serum amylase in dogs and the influence of reticuloendothelial blockade. *Am. J. Physiol.* **210**, 133–8.

Hodges, D. R. & Rhian, M. A. (1962). Surgical procedure for cannulation of thoracic duct and right lymphatic ducts of rhesus monkeys for survival experimentation. *Exp. Med. Surg.* **20**, 258–66.

Hofmann, A. F. & Borgström, B. (1962). Physico-chemical state of lipids in intestinal content during their digestion and absorption. *Fed. Proc.* **21**, 43–50.

Hofmann, A. F. & Small, D. M. (1967). Detergent properties of bile salts: correlation with physiological function. *Ann. Rev. Med.* **18**, 333–76.

Holman, H., Nickel, W. F. & Sleisenger, M. H. (1959). Hypoproteinaemia antedating intestinal lesions and possibly due to excessive protein loss into the intestine. *Am. J. Med.* **27**, 963–75.

Holt, P. R. (1964). Dietary treatment of protein loss in intestinal lymphangiectasia. *Pediatr.* **34**, 629–35.

Howard, J. M., Smith, A. K. & Peters, J. J. (1949). Acute pancreatitis: pathways of enzymes into the blood stream. *Surg.* **26**, 161–6.

Huang, H. S. & Goodman, DeW. S. (1965). Vitamin A and carotenoids. 1. Intestinal absorption and metabolism of ^{14}C-labelled vitamin A alcohol and β-carotene in the rat. *J. biol. Chem.* **240**, 2839–44.

References

Huggins, C. B. & Froehlich, J. P. (1966). High concentration of injected titanium dioxide in abdominal lymph nodes. *J. exp. Med.* **124**; 1099–1106.

Hungerford, G. F. (1959). Effect of adrenalectomy and hydrocortisone on lymph glucose in rats. *Proc. Soc. exp. Biol. Med.* **100**, 754–6.

Hungerford, G. F. & Reinhardt, W. O. (1950). Comparison of effects of sodium pentobarbital or ether-induced anaesthesia on rate of flow and cell content of rat thoracic duct lymph. *Am. J. Physiol.* **160**, 9–14.

Hyatt, R. E. & Smith, J. R. (1954). Mechanism of ascites. A physiologic appraisal. *Am. J. Med.* **16**, 434–48.

Hyatt, R. E., Lawrence, G. H. & Smith, J. R. (1955). Observations on the origin of ascites from experimental hepatic congestion. *J. Lab. Clin. Med.* **45**, 274–80.

Hyde, P. M., Doisy, E. A. Jr., Elliott, W. H. & Doisy, E. A. (1954). Absorption of enterally administered 17 α-methyl-¹⁴C-testosterone and its metabolites. *J. biol. Chem.* **209**, 257–63.

Hyun, S. A., Vahouny, G. V. & Treadwell, C. R. (1967). Portal absorption of fatty acids in lymph and portal vein-cannulated rats. *Biochim. Biophys. Acta* **137**, 296–305.

Inglis, N. R., Ghosh, N. K. & Fishman, W. H. (1968). Sephadex G-200 gel electrophoresis of human serum alkaline phosphatases. *Anal. Biochem.* **22**, 382–6.

Intaglietta, M. (1967). Evidence for a gradient of permeability in frog mesenteric capillaries. *Bibl. Anat.* **9**, 465–8.

Ismail, A. M. & Aboul-Enein, A. (1976). The role of lymphatics in the formation of ascites complicating schistosomal hepatic fibrosis. *Lymphol.* **9**, 43–6.

Isselbacher, K. J. (1966). Biochemical aspects of fat absorption. *Gastroenterol.* **50**, 78–82.

Isselbacher, K. J. & Budz, D. M. (1963). Synthesis of lipoproteins by rat intestinal mucosa. *Nature, Lond.* **200**, 364–5.

Ivey, K., DenBesten, L., Kent, T. H. & Clifton, J. A. (1969). Lymphangiectasia of the colon with protein loss and malabsorption. *Gastroenterol.* **57**, 709–14.

Jackson, R. L., Morrisett, J. D. & Gotto, A. M. (1976). Lipoprotein structure and metabolism. *Phys. Rev.* **56**, 259–316.

Jacobs, F. A. & Largis, E. E. (1969a). Transport of amino acids via the mesenteric lymph duct in rats. *Proc. Soc. exp. Biol. Med.* **130**, 692–6.

Jacobs, F. A. & Largis, E. E. (1969b). Effect of protein inhibitors on protein and amino acids in mesenteric lymph. *Proc. Soc. exp. Biol. Med.* **130**, 697–702.

Jaffe, B. M. & Newton, W. T. (1969). Distribution and localization of radioiodinated gastrin. *Surg. Forum* **20**, 312–13.

James, I. (1975). Prescribing in patients with liver disease. *Br. J. hosp. Med.* **13**, 67–76.

Jaques, L. B. & Waters, E. T. (1941). The identity and origin of the anticoagulant of anaphylactic shock in the dog. *J. Physiol., Lond.* **99**, 454–66.

Jaques, L. B., Millar, G. J. & Spinks, J. W. T. (1954). The metabolism of the K vitamins. *Schweiz Med. Wochenschr.* **84**, 792–6.

Jeejeebhoy, K. N. (1962). Cause of hypoalbuminaemia in patients with gastro-intestinal and cardiac disease. *Lancet* **i** 343–8.

Jeffries, G. H., Chapman, A. & Sleisenger, M. H. (1964). Low fat diet in intestinal lymphangiectasia. *New Eng. J. Med.* **270**, 761–6.

Job, T. T. (1918). Lymphatico-venous communications in the common rat and their significance. *Am. J. Anat.* **24**, 467–91.

Jodal, M. (1973). The significance of the intestinal countercurrent exchanger for the absorption of sodium and fatty acids. Thesis, University of Göteborg, Sweden.

Jodal, M. (1977). The intestinal countercurrent exchanger and its influence on intestinal absorption. In *Intestinal Permeation*, Excerpta Medica, Amsterdam, pp. 48–54.

Jodal, M. & Lundgren, O. (1970). Plasma skimming in the intestinal tract. *Acta physiol. Scand.* **80**, 50–60.

Joel, D. D. & Sautter, J. H. (1963). Preparation of a chronic thoracic duct – venous shunt in calves. *Proc. Soc. exp. Biol. Med.* **112**, 856–9.

Johnson, P. C. (1959). Myogenic nature of increase in intestinal vascular resistance with venous pressure elevation. *Circulation Res.* **7**, 992–9.

Johnson, P. C. (1960). Autoregulation of intestinal blood flow. *Am. J. Physiol.* **199**, 311–18.

Johnson, P. C. & Hanson, K. M. (1962). Effect of arterial pressure on arterial and venous resistance of intestine. *J. appl. Physiol.* **17**, 503–8.

Johnson, P. C. & Hanson, K. M. (1963). Relation between venous pressure and blood volume in the intestine. *Am. J. Physiol.* **204**, 31–4.

Johnson, P. C. & Richardson, D. R. (1974). The influence of venous pressure on filtration forces in the intestine. *Microvasc. Res.* **7**, 296–306.

Johnson, P. & Pover, W. F. R. (1962). Intestinal absorption of α-tocopherol. *Life Sci.* **1**, 115–17.

Johnston, I. D. A. & Code, C. F. (1960). Factors affecting gastric secretion in thoracic-duct lymph of dogs. *Am. J. Physiol.* **198**, 721–4.

Johnston, J. M. (1968). Mechanism of fat absorption. In *Handbook of Physiology* Section 6. *Alimentary Canal, Vol. III Intestinal Absorption.* American Physiological Society, Washington, pp. 1353–75.

Jones, R., Thomas, W. A. & Scott, R. F. (1962). Electron microscopy study of chyle from rats fed butter or corn oil. *Exp. molec. Path.* **1**, 65–83.

Jones, R., Scott, R. F., Morrison, E. S., Kroms, M. & Thomas, W. A. (1963). Biochemical study of lipids in chyle, blood and liver of corn oil- and butter-fed rats with phase and electron microscopy correlation. *Exp. molec. Path.* **2**, 14–31.

Kaihara, S., Nishimura, H., Aoyagi, T., Kameda, H. & Ueda, H. (1963). Protein-losing gastroenteropathy as a cause of hypoproteinaemia in constrictive pericarditis. *Jap. Heart. J.* **4**, 386–94.

Kalima, T. V. (1971). The structure and function of intestinal lymphatics and the influence of impaired lymph flow on the ileum of rats. *Scand. J. Gastroenterol.* **6** Suppl. 10.

Kalima, T. V. & Collan, Y. (1970). Intestinal villus in experimental lymphatic obstruction. Correlation of light and electron microscopic findings with clinical diseases. *Scand. J. Gastroenterol.* **5**, 497–510.

Kamei, Y. (1969). The distribution and relative location of the lymphatic and blood vessels in the mucosa of the rabbit colon. *Nagoya Med. J.* **15**, 223–38.

Kampmeier, O. F. (1960). The development of the jugular lymph sacs in the light of vestigial, provisional and definite phases of morphogenesis. *Am. J. Anat.* **107**, 153–75.

Kampmeier, O. F. (1969). *Evolution and Comparative Morphology of the Lymphatic System.* Charles C. Thomas, Springfield, Illinois.

Kampp, M., Lundgren, O. & Nilsson, N. J. (1967). Extravascular short-circuiting of oxygen indicating countercurrent exchange in the intestinal villi of the cat. *Experientia* **23**, 197–8.

Karmen, A., Whyte, M. & Goodman, D. S. (1963). Fatty acid esterification and chylomicron formation during fat absorption. 1. Triglycerides and cholesterol esters. 2. Phospholipids. *J. Lipid Res.* **4**, 312–29.

Katayama, K. & Fujita, T. (1972a). Studies on lymphatic absorption of $1',2'$-(^{3}H)-Coenzyme Q_{10} in rats. *Chem. Pharm. Bull.* **20**, 2585–92.

Katayama, K. & Fujita, T. (1972b). Studies on biotransformation of elastase. II. Intestinal absorption of ^{131}I-labelled elastase *in vivo*. *Biochim. Biophys. Acta* **288**, 181–9.

Kay, D. & Robinson, D. S. (1962). The structure of chylomicra obtained from the thoracic duct of the rat. *Q. J. exp. Physiol.* **47**, 258–61.

289

References

Kayden, H. J. & Medick, M. (1969). The absorption and metabolism of short and long chain fatty acids in puromycin-treated rats. *Biochim. Biophys. Acta* **176**, 37–43.

Keiding, N. R. (1964). The alkaline phosphatase fractions of human lymph. *Clin. Sci.* **26**, 291–7.

Kelly, K. A., Ikard, R. W., Nyhus, L. M. & Harkins, H. N. (1963). Gastric secretagogues in postprandial thoracic duct lymph. *Am. J. Physiol.* **205**, 85–8.

Kessler, J. I., Stein, J., Dannacker, D. & Narcessian, P. (1970). Biosynthesis of low density lipoprotein by cell-free preparations of rat intestinal mucosa. *J. biol. Chem.* **245**, 5281–8.

Kim, K. S. & Bollman, J. L. (1954). Effect of electrolytes on formation of intestinal lymph in rats. *Am. J. Physiol.* **179**, 273–8.

Kim, M. S. & Ivy, A. C. (1933). On the mode of action of secretagogues (liver extract) in promoting gastric secretion. *Am. J. Physiol.* **105**, 220–40.

Kinmonth, J. B. (1972). *The Lymphatics. Diseases, Lymphography and Surgery.* Williams and Wilkins, Baltimore.

Kinmonth, J. B. & Sharpey-Schafer, E. P. (1959). Pressure waves in the human thoracic duct. *J. Physiol., Lond.* **145**, 3P.

Kinmonth, J. B. & Taylor, G. W. (1956). Spontaneous rhythmic contractility in human lymphatics. *J. Physiol., Lond.* **133**, 3P.

Klatskin, G. & Molander, D. W. (1952). The absorption and excretion of tocopherol in Laennec's cirrhosis. *J. clin. Invest.* **31**, 159–70.

Klein, E. (1882). On the lymphatic system and the minute structure of the salivary glands and pancreas. *Q. J. microscop. Sci., NS* **22**, 154–75.

Kobayashi, Y., Kupelian, J. & Maudsley, D. V. (1969). Release of diamine oxidase by heparin in the rat. *Biochem. Pharmacol.* **18**, 1585–91.

Kokas, E. & Johnston, C. L. (1965). Influence of refined villikinin on motility of intestinal villi. *Am. J. Physiol.* **208**, 1196–202.

Koler, R. D. & Mann, J. D. (1951). Iron content of intestinal lymph of rats. *Proc. Soc. exp. Biol. Med.* **76**, 221–2.

Kolmen, S. N. (1961). Chronic thoracic duct–oesophageal shunt preparation in protein studies. *J. appl. Physiol.* **16**, 369–70.

Kolmen, S. N. & Vita, A. E. (1962). Passage of fibrinogen to circulation via lymphatics. *Am. J. Physiol.* **202**, 671–4.

Korner, P. I., Morris, B. & Courtice, F. C. (1954). An analysis of factors affecting lymph flow and protein composition during gastric absorption of food and fluids and during intravenous infusion. *Aust. J. exp. Biol. med. Sci.* **32**, 301–20.

Kostner, G. M. (1976). Apo B-deficiency (a-β-lipoproteinaemia): a model for studying the lipoprotein metabolism. In: *Lipid Absorption: Biochemical and Clinical Aspects,* ed. K. Rommel & H. Goebell, MTP Press, Lancaster.

Kottakis, G. & Agapitides, N. (1970). Experimental and clinical applications of lymphovenous anastomoses (Abstr.). *Third International Congress of Lymphology (Brussels),* p. 34.

Lam, K. C. & Mistilis, S. P. (1973). Role of intestinal alkaline phosphatase in fat transport. *Aust. J. exp. Biol. Med. Sci.* **51**, 411–16.

Lamanna, C. & Carr, C. J. (1967). The botulinal, tetanal and enterostaphylococcal toxins: a review. *Clin. Pharmacol. Ther.* **8**, 286–332.

Lambert, R. (1965). *Surgery of the Digestive System in the Rat.* Charles C. Thomas, Springfield, Illinois.

Landis, E. M. (1927a). Micro-injection studies of capillary permeability. I. Factors in the production of capillary stasis. *Am. J. Physiol.* **81**, 124–42.

Landis, E. M. (1927b). Micro-injection studies of capillary permeability. II. The relation between capillary pressure and the rate at which fluid passes through the walls of single capillaries. *Am. J. Physiol.* **82**, 217–38.

References

Landis, E. M. (1930). The capillary blood pressure in mammalian mesentery as determined by the micro-injection method. *Am. J. Physiol.* **93**, 353–62.

Landis, E. M. (1964). Heteroporosity of the capillary wall as indicated by cinematographic analysis of the passage of dyes. *Ann. N.Y. Acad. Sci.* **116**, 765–73.

Landis, E. M. & Pappenheimer, J. R. (1963). Exchange of substances through the capillary walls. In *Handbook of Physiology*, Section 2, *Circulation*. Vol. II. Ed. W. F. Hamilton. Chap. 29, pp. 961–1034. American Physiological Society, Washington DC.

Landor, J. H. & Baker, W. K. (1964). Gastric hypersecretion produced by massive small bowel resection in dogs. *J. Surg. Res.* **4**, 518–22.

Largis, E. E. & Jacobs, F. A. (1971). Effects of phorizin on glucose transport into blood and lymph. *Biochim. Biophys. Acta* **225**, 301–7.

Lascelles, A. K. & Morris, B. (1961). Surgical techniques for the collection of lymph from unanaesthetized sheep. *Q. J. exp. Physiol.* **46**, 199–205.

Lascelles, A. K., Hardwick, D. C., Linzell, J. L. & Mepham, T. B. (1964). The transfer of ^{3}H-stearic acid from chylomicra to milk fat in the goat. *Biochem. J.* **92**, 36–42.

Lassen, N. A., Parving, H. H. & Rossing, N. (1974). Filtration as the main mechanism of overall transcapillary protein escape from the plasma. (Editorial.) *Microvasc. Res.* **7**, i–iv.

Laurent, T. C. (1970). In *Capillary Permeability*, ed. C. Crone and N. A. Lassen, pp. 261–77. Munksgaard, Copenhagen.

Lawrence, C. W., Crain, F. D., Lotspeich, F. J. & Krause, R. F. (1966). Absorption, transport and storage of retinyl-15-^{14}C palmitate-9,10-^{3}H in the rat. *J. Lipid Res.* **7**, 226–9.

Leak, L. V. (1976). The structure of lymphatic capillaries in lymph formation. *Fed. Proc.* **35**, 1863–71.

Leak, L. V. & Burke, J. F. (1966). Fine structure of the lymphatic capillary and the adjoining connective tissue area. *Am. J. Anat.* **118**, 785–809.

Leak, L. V. & Burke, J. F. (1968). Ultrastructural studies on the lymphatic anchoring filaments. *J. Cell Biol.* **36**, 129–49.

Leandoer, L. & Lewis, D. H. (1970). The effect of L-norepinephrine on lymph flow in man. *Ann. Surg.* **171**, 257–60.

Leat, W. M. F. & Harrison, F. A. (1974). Origin and formation of lymph lipids in the sheep. *Q. J. exp. Physiol.* **59**, 131–9.

Lee, D. S. & Hashim, S. A. (1966). A new catheter system for cannulation of thoracic duct of the rat. *J. appl. Physiol.* **21**, 1887–8.

Lee, F. C. (1922). The establishment of collateral circulation following ligation of the thoracic duct. *Bull. Johns Hopkins Hosp.* **33**, 21–31.

Lee, J. S. (1961). Flows and pressures in lymphatic and blood vessels of intestine in water absorption. *Am. J. Physiol.* **200**, 979–83.

Lee, J. S. (1963). Role of the mesenteric lymphatic system in water absorption from rat intestine *in vitro*. *Am. J. Physiol.* **204**, 92–6.

Lee, J. S. (1965). Motility, lymphatic contractility and distension pressure in intestinal absorption. *Am. J. Physiol.* **208**, 621–7.

Lee, J. S. (1969a). Role of lymphatic system in water and solute transport from rat intestine *in vitro*. *Q. J. exp. Physiol.* **54**, 311–21.

Lee, J. S. (1969b). A micropuncture study of water transport by dog jejunal villi *in vitro*. *Am. J. Physiol.* **217**, 1528–33.

Lee, J. S. (1971). Contraction of villi and fluid transport in dog jejunal mucosa *in vitro*. *Am. J. Physiol.* **221**, 488–95.

Lee, J. S. (1974). Glucose concentration and hydrostatic pressure in dog jejunal villus lymph. *Am. J. Physiol.* **226**, 675–81.

References

Lee, J. S. & Duncan, K. M. (1968). Lymphatic and venous transport of water from rat jejunum: a vascular perfusion study. *Gastroenterol.* **54**, 559–67.

Lee, J. S. & Silverberg, J. W. (1972). Effect of cholera toxin on fluid absorption and villus lymph pressure in dog jejunal mucosa. *Gastroenterol.* **62**, 993–1000.

Lever, A. F. & Peart, W. S. (1962). Renin and angiotensin-like activity in renal lymph. *J. Physiol., Lond.* **160**, 548–63.

Levick, J. R. & Michel, C. C. (1973). The permeability of individually perfused frog mesenteric capillaries to T-1824 and T-1824-albumin as evidence for a large pore system. *Quart. J. exp. Physiol.* **58**, 67–85.

Levin, A. S. & Jeffay, H. (1964). Metabolism of serum albumin in rats with cirrhosis of the liver. *J. lab. clin. Med.* **63**, 776–83.

Lewis, G. P. (1975). A lymphatic approach to tissue injury. *New Engl. J. Med.* **293**, 287–91.

Linder, E. & Blomstrand, R. (1958). Technic for collection of thoracic duct lymph of man. *Proc. Soc. exp. Biol. Med.* **97**, 653–7.

Lindgren, F. T. & Nicholls, A. V. (1960). Structure and function of human serum lipoprotein. In *The Plasma Proteins*, Vol. II *Biosynthesis, Metabolism, Alterations in Disease.* pp. 1–58. Ed. F. W. Putnam, Academic Press, New York.

Lindsay, F. E. F. (1974). The cisterna chyli as a source of lymph samples in the cat and dog. *Res. vet. Sci.* **17**, 256–8.

Lindsey, C. A. & Wilson, J. D. (1965). Evidence for a contribution by the intestinal wall to the serum cholesterol of the rat. *J. Lipid Res.* **6**, 173–81.

Linscheer, W. G., Malagelada, J. R. & Fishman, W. H. (1971). Diminished oleic acid absorption in man by L-phenylalanine inhibition of an intestinal phosphohydrolase. *Nature New Biol.* **231**, 116–17.

Logan, G. B. (1961). Release of a histamine-destroying factor during anaphylactic shock in guinea pigs. *Proc. Soc. exp. Biol. Med.* **107**, 466–9.

Lord, R. S. A. (1968). The white veins: conceptual difficulties in the history of the lymphatics. *Med. Hist.* **12**, 174–84.

Ludwig, C. (1858). *Lehrbuch der Physiologie des Menschen.* Winter, Leipzig.

Ludwig, J., Linhart, P. & Baggenstoss, A. H. (1968). Hepatic lymph drainage in cirrhosis and congestive heart failure. *Arch. Pathol.* **86**, 551–62.

Lundgren, O. (1967). Studies on blood flow distribution and countercurrent exchange in the small intestine. *Acta physiol. Scand. Suppl.* **303**, 1–42.

Macallum, A. B. (1895). On the distribution of assimilated iron compounds, other than haemoglobin and haematins, in animal and vegetable cells. *Proc. Roy. Soc. Lond.* **57**, 261.

McCarrell, J. D., Thayer, S. & Drinker, C. K. (1941). The lymph drainage of the gall bladder together with observations on the composition of liver lymph. *Am. J. Physiol.* **133**, 79–81.

McGuigan, J. E., Purkerson, M. L., Trudeau, W. L. & Peterson, M. L. (1968). Studies of the immunologic defects associated with intestinal lymphangiectasia. *Ann. int. Med.* **68**, 398–404.

McGuigan, J. E., Jaffe, B. M. & Newton, W. T. (1970). Immunochemical measurements of endogenous gastrin release. *Gastroenterol.* **59**, 499–504.

McGuigan, J. E., Landor, J. H. & Wickbom, G. (1974). Changes in acid secretion, gastrin release and gastrin metabolism after small intestine resection. *Gastroenterol.* **66**, 742.

McHale, N. G. & Roddie, I. C. (1976). The effect of transmural pressure on pumping activity in isolated bovine lymphatic vessels. *J. Physiol., Lond.* **261**, 255–69.

MacIntosh, F. C. & Krueger, L. (1938). Choline as a stimulant of gastric secretion. *Am. J. Physiol.* **122**, 119–31.

McKee, F. W., Schilling, J. A., Tishkoff, G. H. & Hyatt, R. E. (1949). Experimental ascites; effects of sodium chloride and protein intake on protein metabolism of dogs with constricted inferior vena cava. *Surg. Gynec. Obstet.* **89**, 529–40.

McKee, F. W., Wilt, W. G., Hyatt, R. E. & Whipple, G. H. (1950). The circulation of ascitic fluid. Interchange of plasma and ascitic fluid protein as studied by means of ^{14}C-labelled lysine in dogs with constriction of the vena cava. *J. exp. Med.* **91**, 115–22.

McMahon, M. T. & Neale, G. (1969). Protein losing enteropathy possibly due to filariasis. *Proc. Roy. Soc. Med.* **62**, 1043–4.

MacMahon, M. T. & Thompson, G. R. (1970). Comparison of the absorption of a polar lipid, oleic acid, and a non-polar lipid, α-tocopherol from mixed micellar solutions and emulsions. *Europ. J. clin. Invest.* **1**, 161–6.

MacMahon, M. T., Neale, G. & Thompson, G. R. (1971). Lymphatic and portal venous transport of α-tocopherol and cholesterol. *Europ. J. clin. Invest.* **1**, 288–94.

McMaster, P. D. (1946). The pressure and interstitial resistance prevailing in the normal and edematous skin of animals and man. *J. exp. Med.* **84**, 473–94.

Madsen, N. B. & Tuba, J. (1952). On the source of the alkaline phosphatase in rat serum. *J. biol. Chem.* **195**, 741–50.

Mahadevan, S., Seshadri Sastry, P. & Ganguly, J. (1963). Studies on metabolism of vitamin A. 4. Studies on the mode of absorption of Vitamin A by rat intestine *in vitro*. *Biochem. J.* **88**, 534–9.

Maizel, H., Ruffin, J. M. & Dobbins, W. O. (1970). Whipple's Disease. A review of 19 patients from one hospital and a review of the literature since 1950. *Med.* **49**, 175–205.

Mall, F. (1896). The vessels and walls of the dog's stomach. *Johns Hopkins Hosp. Rep.* **1**, 1–36.

Mallet-Guy, P., Devic, G., Feroldi, J. & Desjaques, P. (1954). Etude expérimentale des ascites; sténoses veineuses post-hépatiques et transposition du foie dans le thorax. *Lyon Chir.* **49**, 153–72.

Mallet-Guy, P., Michoulier, J., Baev, S., Oleskiewicz, L. & Woszczyk, M. (1962). Experimental research on lymphatic circulation of the liver. 1. Immediate data on biliolymphatic permeability. *Lyon Chir.* **58**, 847–59.

Mallet-Guy, P., Michoulier, J. & Baev, S. (1965). In *The Biliary System*, ed. W. Taylor, F. A. Davis, Philadelphia.

Mandel, M. A. (1967). Isolation of mouse lymphocytes for immunologic studies by thoracic duct cannulation. *Proc. Soc. exp. Biol. Med.* **126**, 521–4.

Mann, J. D. & Higgins, G. M. (1950). Lymphocytes in thoracic duct, intestinal and hepatic lymph. *Blood* **5**, 177–90.

Mann, J. D., Mann, F. D. & Bollman, J. L. (1949). Hypoprothrombinaemia due to loss of intestinal lymph. *Am. J. Physiol.* **158**, 311–14.

Mann, T. (1924). *The Magic Mountain*, Transl. H. T. Lowe-Porter. Penguin Mod. Class. Harmondsworth, Middx.

Marshall, W. H., Neyazaki, T. & Abrams, H. L. (1965). Abnormal protein loss after thoracic-duct ligation in dogs. *New Eng. J. Med.* **273**, 1092–4.

Mascagni, P. (1787). *Vasorum Lymphaticorum Corporis Humani Historia et Ichnographia*. Carli, Sienna.

Mattson, F. H. & Volpenhein, R. A. (1964). The digestion and absorption of triglycerides. *J. biol. Chem.* **239**, 2772–7.

Mawhinney, H. J. D. & Roddie, I. C. (1973). Spontaneous activity in isolated bovine mesenteric lymphatics. *J. Physiol., Lond.* **229**, 339–48.

May, A. J. & Whaler, B. C. (1958). The absorption of *Clostridium botulinum* type A toxin from the alimentary canal. *Br. J. exp. Path.* **39**, 307–16.

Mayerson, H. S. (1963). The physiologic importance of lymph. In *Handbook of*

References

Physiology, Section 2, *Circulation* Vol. II, Ed. W. F. Hamilton pp. 1035–73, American Physiological Society, Washington, DC.

Mayerson, H. S., Wolfram, C. G., Shirley, H. H. & Wasserman, K. (1960). Regional differences in capillary permeability. *Am. J. Physiol.* **198**, 155–60.

Mayerson, H. S., Patterson, R. M. McKee, A., LeBrie, S. J. & Mayerson, P. (1962). Permeability of lymphatic vessels. *Am. J. Physiol.* **203**, 98–106.

Mayo, C. & Greene, C. H. (1929). Studies in metabolism of bile; role of lymphatics in early stages of development of obstructive jaundice. *Am. J. Physiol.* **89**, 280–8.

Meyer, J. H., Stevenson, E. A. & Watts, H. D. (1976). The potential role of protein in the absorption of fat. *Gastroenterol.* **70**, 232–9.

Mezick, J. A., Tompkins, R. K. & Cornwell, D. G. (1968). Absorption and intestinal lymphatic transport of ^{14}C-menadione. *Life Sci.* **7**, 153–8.

Milhaud, G. & Vesin, P. (1961). Calcium metabolism in man with calcium[45]: Malabsorption syndrome with exudative enteropathy. *Nature, Lond.* **191**, 872–4.

Minari, O. & Zilversmit, D. B. (1963). Behaviour of dog lymph chylomicron lipid constituents during incubation with serum. *J. Lipid Res.* **4**, 424–36.

Mislin, H. (1961). Experimental detection of autochthonous automatism of lymph vessels. *Experientia* **17**, 29–30.

Mislin, H. (1976). Active contractility of the lymphangion and coordination of lymphangion chains. *Experientia*, **32**, 820–2.

Mistilis, S. P. & Ockner, R. K. (1972). Effects of ethanol on endogenous lipid and lipoprotein metabolism in small intestine. *J. Lab. clin. Med.* **80**, 34–46.

Mistilis, S. P. & Skyring, A. P. (1966). Intestinal lymphangiectasia. Therapeutic effect of lymph venous anastomosis. *Am. J. Med.* **40**, 634–41.

Mistilis, S. P., Skyring, A. P. & Stephen, D. D. (1965). Intestinal lymphangiectasia. Mechanism of enteric loss of plasma-protein and fat. *Lancet* **i**, 77–80.

Mizuno, Y., Yasuga, N., Sawada, K., Matsumoto, I., Iwasaki, M., Kayashima, Y., Kimura, K. & Kajiyama, Y. (1966). Pancreatic damage due to lymph congestion: a cause of malabsorption in the patients with intestinal lymphangiectasia. *Acta Gastroenterol. Belg.* **29**, 248–59.

Mohiuddin, A. (1966). Blood and lymph vessels in the jejunal villi of the white rat. *Anat. Rec.* **156**, 83–90.

Molnár, Z., Karoliny, G., Mozsik, Gy. & Németh, A. (1974). Lymphogenic enteropathy from thoracic duct obstruction. *Lymphol.* **3**, 130–6.

Monkhouse, F. C., Fidlar, E. & Barlow, J. C. D. (1952). Release of heparin in anaphylactic shock in irradiated and non-irradiated animals. *Am. J. Physiol.* **169**, 712–20.

Moreno, A. H., Ruzicka, F. F., Rousselot, L. M., Burchell, A. R., Bono, R. F., Slavsky, S. F. & Burke, J. H. (1963). Functional hepatography. A study of the hemodynamics of the outflow tracts of the human liver by intraparenchymal deposition of contrast medium, with attempts at functional evaluation of the outflow block concept of cirrhotic ascites and the accessory outflow role of the portal vein. *Radiology* **81**, 65–79.

Morgan, E. H. (1963). Exchange of iron and transferrin across endothelial surfaces in the rat and rabbit. *J. Physiol., Lond.* **169**, 339–52.

Morgan, R. G. H. & Borgström, B. (1969). The mechanism of fat absorption in the bile fistula rat. *Q. J. exp. Physiol.* **54**, 228–43.

Morris, B. (1956a). The hepatic and intestinal contributions to the thoracic duct lymph. *Q. J. exp. Physiol.* **41**, 318–25.

Morris, B. (1956b). The exchange of protein between the plasma and the liver and intestinal lymph. *Q. J. exp. Physiol.* **41**, 326–40.

Morris, B. & Sass, M. B. (1966). The formation of lymph in the ovary. *Proc. R. Soc.* **164B**, 577–91.

Mortimore, G. E. & Tietze, F. (1959). Studies on the mechanism of capture and degradation of insulin-^{131}I by the cyclically perfused rat liver. *Ann. N.Y. Acad. Sci.* **82**, 329–37.

Mueller, J. H. (1915). The assimilation of cholesterol and its esters. *J. biol. Chem.* **22**, 1–9.

Munk, I. & Rosenstein, A. (1891). Zur Lehre von der Resorption im Darm, nach Untersuchungen an einer Lymph (chylus) fistel beim Menschen. *Virchows Arch. Path. Anat. Physiol.* **123**, 230–79, 484–518.

Muralidhara, K. S. & Hollander, D. (1977). Intestinal absorption of α-tocopherol in the unanaesthetized rat. The influence of luminal constituents on the absorptive process. *J. Lab. Clin. Med.* **90**, 85–91.

Murray, T. K. & Grice, H. C. (1961). Absorption of Vitamin A after thoracic duct ligation in the rat. *Can. J. Biochem.* **39**, 1103–6.

Nelson, A. W. & Swan, H. (1969). Long-term catheterization of the thoracic duct in the dog. *Arch. Surg., Chicago* **98**, 83–6.

Nesis, L. & Sterns, E. E. (1973). Lymph flow from the colon under varying conditions. *Ann. Surg.* **177**, 422–7.

Neyazaki, T., Kupic, E. A., Marshall, W. H. & Abrams, H. L. (1965). Collateral lymphatico-venous communications after experimental obstruction of the thoracic duct. *Radiology* **85**, 423–32.

Nielsen, A. E. (1942). A translation of Olaf Rudbeck's *Nova Exercitatio Anatomica* (1653). *Bull. Hist. Med.* **11**, 311–39.

Nilsson, Å. (1968a). Intestinal absorption of lecithin and lysolecithin by lymph fistula rats. *Biochim. Biophys. Acta* **152**, 379–90.

Nilsson, Å. (1968b). Metabolism of sphingomyelin in the intestinal tract of the rat. *Biochim. Biophys. Acta* **164**, 575–84.

Nilsson, Å. (1969a). The presence of sphingomyelin- and ceramide-cleaving enzymes in the small intestinal tract. *Biochim. Biophys. Acta* **176**, 339–47.

Nilsson, Å. (1969b). Intestinal absorption and metabolism of lecithin, sphingomyelin and cerebroside. Thesis, University of Lund, Sweden.

Nix, J. T., Flock, E. V. & Bollman, J. L. (1951). Influence of cirrhosis on proteins of cisternal lymph. *Am. J. Physiol.* **164**, 117–18.

Nix, J. T., Mann, F. C., Bollman, J. L., Grindlay, J. H. & Flock, E. V. (1951). Alterations of protein constituents of lymph by specific injury to the liver. *Am. J. Physiol.* **164**, 119–22.

Norman, A. W., Mircheff, A. K., Adams, T. H. & Spielvogel, A. (1970). Studies on the mechanism of action of calciferol. III. Vitamin D-mediated increase of intestinal brush border alkaline phosphatase activity. *Biochim. Biophys. Acta* **215**, 348–59.

Nothacker, W. G. & Brauer, R. W. (1950). Histopathology of the liver following histamine infusion. *Proc. Soc. exp. Biol. Med.* **75**, 749–52.

Noyan, A. (1964). Water absorption from the intestine via portal and lymphatic pathways in rats. *Proc. Soc. exp. Biol. Med.* **117**, 317–20.

Nusbaum, M., Baum, S., Rajatapiti, B. & Blakemore, W. S. (1967). Intestinal lymphangiography in vivo. *J. Cardiovasc. Surg.* **8**, 62–8.

Nylander, G. & Tjernberg, B. (1969). The lymphatics of the greater omentum. An experimental study in the dog. *Lymphol.* **2**, 3–7.

Ockner, R. K., Hughes, F. B. & Isselbacher, K. J. (1969a). Very low density lipoproteins in intestinal lymph: origin, composition and role in lipid transport in the fasting state. *J. clin. Invest.* **48**, 2079–88.

Ockner, R. K., Hughes, F. B. & Isselbacher, K. J. (1969b). Very low density lipoproteins in intestinal lymph: role in triglyceride and cholesterol transport during fat absorption. *J. clin. Invest.* **48**, 2367–73.

Ockner, R. K., Manning, J. A., Poppenhausen, R. B. & Ho, W. K. L. (1972). A

References

binding protein for fatty acids in cytosol of intestinal mucosa, liver, myocardium and other tissues. *Science* **177**, 56–8.

Ockner, R. K., Mistilis, S. P., Poppenhausen, R. B. & Stiehl, A. F. (1973). Ethanol-induced fatty liver: effect of intestinal lymph fistula. *Gastroenterol.* **64**, 603–9.

O'Doherty, P. J. A., Kakis, G. & Kuksis, A. (1973). Role of luminal lecithin in intestinal fat absorption. *Lipids* **8**, 249–55.

Ohlsson, K. (1971). Experimental pancreatitis in the dog. Appearance of complexes between proteases and trypsin inhibitors in ascitic fluid, lymph and plasma. *Scand. J. Gastroenterol.* **6**, 645–52.

Olarte, J. A., Sachetto, J. R., Vilarino, J., Rubio, H. H. & Fernandez, F. J. (1964). Faecal excretion of PVP-I^{131} in liver cirrhosis. *Gastroenterologia* **101**, 137–44.

Oliver, G. C., Cooksey, J., Witte, C. & Witte, M. (1971). Absorption and transport of digitoxin in the dog. *Circ. Res.* **29**, 419–23.

Olson, J. A. (1961). The conversion of radioactive β-carotene into vitamin A by the rat intestine *in vivo. J. biol. Chem.* **236**, 349–56.

O'Morchoe, C. C. C. & O'Morchoe, P. J. (1968). Renal contribution to thoracic duct lymph in dogs. *J. Physiol., Lond.* **194**, 305–15.

Orloff, M. J., Wright, P. W., DeBenedetti, M. J., Halasz, N. A., Annetts, D. L., Musicant, M. E. & Goodhead, B. (1966). Experimental ascites. VII. The effects of external drainage of the thoracic duct on ascites and hepatic hemodynamics. *Arch. Surg.* **93**, 119–30.

Orloff, M. J., Goodhead, B., Windsor, C. W. O., Musicant, M. E. & Annetts, D. L. (1967). Effect of portacaval shunts on lymph flow in the thoracic duct. *Am. J. Surgery* **114**, 213–21.

Osato, S. (1921). Beiträge zum Studium der Lymphe. *Tohoku J. exp. Med.* **2**, 465–530.

Ottaviani, G. & Azzali, G. (1969). Ultrastructure of lymphatic vessels in some functional conditions. *Acta Anat. (Basle)* **73, Suppl. 56**, 325–36.

Palay, S. L. & Karlin, L. J. (1959a). An electron microscopic study of the intestinal villus. 1. The fasting animal. *J. Biophys. Biochem. Cytol.* **5**, 363–72.

Palay, S. L. & Karlin, L. J. (1959b). An electron microscopic study of the intestinal villus. 2. The pathway of fat absorption. *J. Biophys. Biochem. Cytol.* **5**, 373–84.

Paldino, R. L. & Hyman, C. (1964). Relationship between lymphatic and blood flow in various structures in the abdominal cavity. *Proc. Soc. exp. Biol. Med.* **117**, 904–10.

Palla, J. C., Ben Abdeljlil, A. & Desnuelle, P. (1968). Action de l'insuline sur la biosynthèse de l'amylase et de quelques autres enzymes du pancréas de rat. *Biochim. Biophys. Acta* **158**, 25–35.

Panizza, B. (1833). *Sopra il Sistema Linfatico dei rettili Ricerche Zootomiche. . .con sei tavole incise in rame.* P. Bizzoni, Pavia.

Papp, M. & Fodor, I. (1957). Lymphatic system of the salivary glands. (Cited by Rusznyak, I., Földi, M. & Szabo, G. (1967). In *Lymphatics and Lymph Circulation*, 2nd ed. Pergamon Press, Oxford.)

Papp, M., Röhlich, P., Rusznyák, I. & Toro, I. (1962). An electron microscopic study of the central lacteal in the intestinal villus of the cat. *Zeit. Zellforsch.* **57**, 475–86.

Papp, M., Makara, G. B. & Folly, G. (1975). Impeded interstitial fluid movement: a factor in pancreatic oedema. *Lymphol.* **8**, 148–53.

Papp, M., Németh, E., Feuer, I. & Fodor, I. (1958). Effect of an impairment of lymph flow on experimental acute pancreatitis. *Acta Med. Acad. Sci. Hung.* **11**, 203–8.

Papp, M., Németh, E. P. & Horváth, E. J. (1971). Pancreatico-duodenal lymph flow and lipase activity in acute experimental pancreatitis. *Lymphol.* **4**, 48–53.

Papp, M., Ormai, S., Horváth, E. J. & Fodor, I. (1971). The effect of secretin and pancreozymin on pancreatico-duodenal lymph flow and lipase activity in normal dogs and on thoracic duct lymph flow and lipase activity in rats with chronic pancreatitis. *Lymphol.* **4**, 67–73.

Papp, M., Varga, B. & Makara, G. B. (1973). The role of lymph nodes in pancreatic oedema. *Lymphol.* **6**, 28–34.

Pappenheimer, J. R. (1953). The passage of molecules through capillary walls. *Physiol. Rev.* **33**, 387–423.

Pappenheimer, J. R. & Soto-Rivera, A. (1948). Effective osmotic pressure of the plasma proteins and other quantities associated with the capillary circulation in the hind limbs of cats and dogs. *Am. J. Physiol.* **152**, 471–91.

Patterson, R. M., Ballard, C. L., Wasserman, K. & Mayerson, H. S. (1958). Lymphatic permeability to albumin. *Am. J. Physiol.* **194**, 120–4.

Pecquet, J. (1653). *New Anatomical Experiments by which the hitherto unknown receptacle of the Chyle and the transmission from thence to the Subclavial veins by the now discovered Lacteal Channels of the Thorax is plainly made appear in Brutes.* T. W. Pulleyn, London.

Pepin, J., Singh, H., Pairent, F. W., Appert, H. E. & Howard, J. M. (1970). A study of insulin secretion in thoracic duct lymph of the dog. *Ann. Surg.* **172**, 56–60.

Perry, M. & Garlick, D. (1975). Transcapillary efflux of gamma globulin in rabbit skeletal muscle. *Microvasc. Res.* **9**, 119–26.

Perry, T. T. (1947). Role of lymphatic vessels in transmission of lipase in disseminated pancreatic fat necrosis. *Arch. Path.* **43**, 456–65.

Peters, T. J. & MacMahon, M. T. (1970). The absorption of glycine and glycine oligopeptides by the rat. *Clin. Sci.* **39**, 811–21.

Petersen, V. P. & Hastrup, J. (1963). Protein-losing enteropathy in constrictive pericarditis. *Acta Med. Scand.* **173**, 401–10.

Petersen, V. P. & Ottosen, P. (1964). Albumin turnover and thoracic duct lymph in constrictive pericarditis. *Acta Med. Scand.* **176**, 335–44.

Peterson, R. E. & Mann, J. D. (1952). Transport of radioactive iron in intestinal lymph. *Am. J. Physiol.* **169**, 763–6.

Pflug, J. & Calnan, J. (1968). The valves of thoracic duct at the angulus venosus. *Brit. J. Surg.* **55**, 911–16.

Pick, J. W., Anson, B. J. & Burnett, H. W. (1964). Communication between lymphatic and venous sytem at the renal level in man. *Northwestern Univ. Med. School Q. Bull.* **18**, 307.

Pierce, A. E., Risdall, P. C. & Shaw, B. (1964). Absorption of orally administered insulin by the newly born calf. *J. Physiol., Lond.* **171**, 203–15.

Pihl, O., Iber, F. L. & Linscheer, W. G. (1970). The enhancement of vitamin D_3 absorption in man by medium and long chain fatty acids. *Clin. Res.* **18**, 462.

Pinter, G. G. & Zilversmit, D. B. (1962). A gradient centrifugation method for the determination of particle size distribution of chylomicrons and of fat droplets in artificial fat emulsions. *Biochim. Biophys. Acta* **59**, 116–27.

Pirola, R. C. & Davis, A. E. (1970). Effect of pressure on the integrity of the duct-acinar system of the pancreas. *Gut* **11**, 69–73.

Planche, N. E., Hage, G., Montet, A. M. & Sarles, H. (1975). Experimental lymphatic obstruction and fat absorption in the rabbit. *Digestion.* **13**, 255–58.

Plauth, W. H., Waldmann, T. A., Wochner, R. D., Braunwald, N. S. & Braunwald, E. (1964). Protein-losing enteropathy secondary to constrictive pericarditis in childhood. *Pediat.* **34**, 636–48.

Playoust, M. R. & Isselbacher, K. J. (1964). Studies on the intestinal absorption and intramucosal lipolysis of a medium chain triglyceride. *J. clin. Invest.* **43**, 878–85.

Polderman, H., McCarrell, J. D. & Beecher, H. K. (1943). Effect of anaesthesia on lymph flow (local procaine, ether and pentobarbital sodium). *J. Pharmac. exp. Ther.* **78**, 400–6.

References

Pollard, T. D. & Weihing, R. R. (1974). Actin and myosin and cell movement. *C.R.C. Crit. Rev. Biochem.* **2**, 1–65.

Pomerantz, M. & Waldmann, T. A. (1963). Systemic lymphatic abnormalities associated with gastro-intestinal protein loss secondary to intestinal lymphangiectasia. *Gastroenterol.* **45**, 703–11.

Popper, H. & Volk, B. W. (1944). Absorption of vitamin A in the rat. *Arch. Path.* **38**, 71–5.

Popper, H. L. & Necheles, H. (1940). Pathways of enzymes into the blood in acute damage of the pancreas. *Proc. Soc. exp. Biol. Med.* **43**, 220–2.

Pressman, J. J. & Simon, M. B. (1961). Experimental evidence of direct communications between lymph nodes and veins. *Surg. Gynec. Obstet.* **113**, 537–41.

Pressman, J. J., Simon, M. B., Hand, K. & Miller, J. (1962). Passage of fluids, cells and bacteria via direct communications between lymph nodes and veins. *Surg. Gynec. Obstet.* **115**, 207–14.

Pressman, J. J., Burtz, M. V. & Shafer, L. (1964). Further observations related to direct communications between lymph nodes and veins. *Surg. Gynec. Obstet.* **119**, 984–90.

Pullinger, B. D. & Florey, H. W. (1935). Some observations on the structure and function of lymphatics: their behaviour in local oedema. *Br. J. exp. Path.* **16**, 49–61.

Quin, J. W. & Lascelles, A. K. (1975). Relationship between the recirculation of lymphocytes and protein concentration of lymph in sheep. *Aust. J. exp. Biol. Med. Sci.* **53**, 1–9.

Quin, J. W. & Shannon, A. D. (1975). The effect of anaesthesia and surgery on lymph flow, protein and leucocyte concentration in lymph of the sheep. *Lymphol.* **8**, 126–35.

Quin, J. W., Husband, A. J. & Lascelles, A. K. (1975). The origin of the immunoglobulins in intestinal lymph of sheep. *Aust. J. exp. Biol. med. Sci.* **53**, 205–14.

Ramos, O. L., Saad, F. & Leser, W. P. (1964). Portal hemodynamics and liver cell function in hepatic schistosomiasis. *Gastroenterol.* **47**, 241–7.

Rampone, A. J. (1959). Experimental thoracic duct fistula for conscious dogs. *J. appl. Physiol.* **14**, 150–2.

Ranvier, L. (1896). Des lymphatiques de la villosité intestinale chez le rat et le lapin. *C.R. Séanc. Acad. Sci. Paris* **123**, 923–5.

Rasio, E. A., Soeldner, J. S. & Cahill, G. F. (1965). Insulin and insulin-like activity in serum and extravascular fluid. *Diabetol.* **1**, 125–7.

Rasio, E. A., Hampers, C. L., Soeldner, J. S. & Cahill, G. F. (1967). Diffusion of glucose, insulin, inulin and Evans blue protein into thoracic duct lymph of man. *J. clin. Invest.* **46**, 903–10.

Raybuck, H. E., Weatherford, T. & Allen, L. (1960). Lymphatics in genesis of ascites in the rat. *Am. J. Physiol.* **198**, 1207–10.

Redgrave, T. G. (1969). Inhibition of protein synthesis and absorption of lipid into thoracic duct lymph of rats. *Proc. Soc. exp. Biol. Med.* **130**, 776–80.

Redgrave, T. G. & Dunne, K. B. (1975). Chylomicron formation and composition in unanaesthetised rabbits. *Atheroscler.* **22**, 389–400.

Redgrave, T. G. & Zilversmit, D. B. (1969). Does puromycin block release of chylomicrons from the intestine? *Am. J. Physiol.* **217**, 336–40.

Reichert, F. L. & Mathes, M. E. (1936). Experimental lymphedema of the intestinal tract and its relation to regional cicatrizing enteritis. *Ann. Surg.* **104**, 601–16.

Reininger, E. J. & Sapirstein, L. A. (1957). Effect of digestion on distribution of blood flow in the rat. *Science* **126**, 1176.

Reinke, R. T. & Wilson, J. D. (1967). The mechanism of bile acid absorption into the portal circulation. *Clin. Res.* **15**, 242.

Reiser, R. & Bryson, M. J. (1951). Route of absorption of free fatty acids and triglycerides from the intestine. *J. biol. Chem.* **189**, 87–91.

References

Reizenstein, P. G., Cronkite, E. P., Meyer, L. M. & Usenik, E. A. (1960). Lymphatics in intestinal absorption of vitamin B_{12} and iron. *Proc. Soc. exp. Biol. Med.* **105**, 233–6.

Rényi-Vámos, F. & Szinay, G. (1954). Das Lymphgefässsystem des Magens und sein Verhalten bei Ulcus ventriculi. *Acta morph. Acad. sci. Hung.* **4**, 353–65.

Retik, A. B., Perlmutter, A. D. & Harrison, J. H. (1965). Communications between lymphatics and veins involving the portal circulation. *Am. J. Surg.* **109**, 201–5.

Reynolds, B. L. (1967). Stimulation of secreting cell mass by submucosal lymph from gastric antrum in dogs. *Am. Surg.* **33**, 352–8.

Reynolds, B. M. (1970). Observations of the subcapsular lymphatics in normal and diseased human pancreas. *Ann. Surg.* **171**, 559–66.

Reza Rafii, M. (1972). Mesenteriale Lymphgefäss veränderungen bei Leberzirrhose und Rechtsherz Insuffizienz. Morphologische Beiträge. *Fortschr. Med.* **90**, 1295–8.

Rhodin, J. A. G. (1974). *Histology. A Text and Atlas*. Oxford University Press. New York.

Rich, A. R. & Duff, G. L. (1936). Experimental and pathological studies on the pathogenesis of acute hemorrhagic pancreatitis. *Bull. Johns Hopkins Hosp.* **58**, 212–59.

Ritchie, H. D., Grindlay, J. H. & Bollman, J. L. (1956). Surgical jaundice: Experimental evidence against the 'regurgitation theory'. *Surg. Forum* **7**, 415–18.

Ritchie, H. D., Grindlay, J. H. & Bollman, J. L. (1959). Flow of lymph from the canine liver. *Am. J. Physiol.* **196**, 105–9.

Roberts, S. H. & Douglas, A. P. (1976). Intestinal lymphangiectasia: the variability of presentation. A study of five cases. *Q. J. Med. New Series* **45**, 39–48.

Robinson, D. S. (1955). Chemical composition of chylomicra in the rat. *Q. J. exp. Physiol.* **40**, 112–26.

Roddenberry, H. & Allen, L. (1967). Observations on the abdominal lymphatico-venous communications of the squirrel monkey (*Saimiri sciureus*). *Anat. Rec.* **159**, 147–57.

Rosenthal, W. S. & Rudick, J. (1971). Chylogastrone: gastric secretory inhibitor in thoracic duct lymph of man and dog. *Am. J. Physiol.* **220**, 452–6.

Ross, G. (1970a). Cardiovascular effects of secretin. *Am. J. Physiol.* **218**, 1166–70.

Ross, G. (1970b). Regional circulatory effects of pancreatic glucagon. *Br. J. Pharmacol.* **38**, 735–42.

Ross, G. (1971). Effects of norepinephrine infusions on mesenteric arterial bloodflow and its tissue distribution. *Proc. Soc. exp. Biol. Med.* **137**, 921–4.

Rous, P., Gilding, H. P. & Smith, F. (1930). The gradient of vascular permeability. *J. exp. Med.* **51**, 807–30.

Rudbeck, O. (1653). *Nova Exercitatio Anatomica Exhibens Ductus Hepaticos Aquosos et Vasa Glandularum Serosa*. Lauringer, Västerås.

Rudick, J., Gajewski, A. K., Pitts, C. L., Semb, L. S., Fletcher, T. L., Harkins, H. N. & Nyhus, L. M. (1965). Gastric inhibitors in fasting canine thoracic duct lymph. *Proc. Soc. exp. Biol. Med.* **120**, 119–21.

Rudick, J., Gajewski, A. K., Pitts, C. L., Semb, L. S., Fletcher, T. L., Harkins, H. N. & Nyhus, L. M. (1966). Inhibition of gastric secretion by fractionated lymph. Elaboration of gastrone-like substance during digestive phases in dogs. *Am. Surg.* **32**, 513–20.

Rudick, J., Fletcher, T. L. & Dreiling, D. A. (1968). Effects of lymph diversion on the gastric secretory response to endogenous gastrin. *Am. J. Physiol.* **215**, 370–4.

Rudick, J., Fletcher, T. L., Dreiling, D. A. & Kark, A. E. (1968). Influence of lymph diversion on the gastric secretory response to feeding. *Proc. Soc. exp. Biol. Med.* **128**, 895–7.

Rusznyak, I., Földi, M. & Szabo, G. (1967). *Lymphatics and Lymph Circulation. Physiology and Pathology*. Pergamon Press, Oxford.

Rutili, G. & Arfors, K. E. (1977). Protein concentration in interstitial and lymphatic fluids from the subcutaneous tissue. *Acta Physiol. Scand.* **99**, 1–8.

References

Ruysch, F. (1665). *Dilucidatio Valvularum in Vasis Lymphaticis et Lacteis.* H. Gael, The Hague.

Sabesin, S. M. (1976). Ultrastructural aspects of the intracellular assembly, transport and exocytosis of chylomicrons by rat intestinal absorptive cells. In *Lipid Absorption: Biochemical and Clinical Aspects*, ed. K. Rommel & H. Goebell. MTP Press Lancaster.

Sabesin, S. M. & Isselbacher, K. J. (1965). Protein synthesis inhibition: mechanism for the production of impaired fat absorption. *Science* **147**, 1149–51.

Sabesin, S. M., Holt, P. R. & Clark, S. B. (1975). Intestinal lipid absorption: evidence for an intrinsic defect of chylomicron secretion by normal rat distal intestine. *Lipids* **10**, 840–6.

Sabin, F. R. (1916). The origin and development of the lymphatic system. *Johns Hopkins Hosp. Rep.* **17**, 347–440.

Sadek, A. M., Aboul-Enein, A., Hassenein, E. & Ismail, A. (1970). Thoracic duct changes in schistosomal hepatic fibrosis. *Gut* **11**, 74–8.

Said, S. I. & Mutt, V. (1970). Potent peripheral and splanchnic vasodilator peptide from normal gut. *Nature, Lond.* **225**, 863–4.

Saini, P. K. & Posen, S. (1969). The origin of serum alkaline phosphatase in the rat. *Biochim. Biophys. Acta* **177**, 42–9.

Salpeter, M. M. & Zilversmit, D. B. (1968). The surface coat of chylomicrons: electron microscopy. *J. Lipid Res.* **9**, 187–92.

Samanta, A. K. S., Saini, V. K., Chhuttani, P. N., Patra, B. S., Vashista, S. & Datta, D. V. (1974). Thoracic duct and hepatic lymph in idiopathic portal hypertension. *Gut* **15**, 903–6.

Sarda, L., Maylié, M. F., Roger, J. & Desnuelle, P. (1964). Comportement de la lipase pancréatique sur Sephadex. Application à la purification et à la determination du poids moléculaire de cet enzyme. *Biochim. Biophys. Acta* **89**, 183–5.

Saunders, D. R. & Dawson, A. M. (1963). The absorption of oleic acid in the bile fistula rat. *Gut* **4**, 254–60.

Saunders, D. R., Parmentier, C. M. & Ways, P. O. (1968). Metabolism of lysolecithin by rat small intestine. *Gastroenterol.* **54**, 382–91.

Saunders, W. (1795). Cited by Mayo, C. & Greene, C. H. (1929). *Am. J. Physiol.* **89**, 280–88.

Scanu, A. M., Aggerbeck, L. P., Kruski, A. W., Lim, C. T. & Kayden, H. J. (1974). A study of the abnormal lipoproteins in abetalipoproteinemia. *J. clin. Invest.* **53**, 440–53.

Schachter, D., Finkelstein, J. D. & Kowarski, S. (1963). Pathways of transport and metabolism of C^{14}-vitamin D$_2$ in the rat. *J. clin. Invest.* **42**, 974–5.

Schachter, D., Finkelstein, J. D. & Kowarski, S. (1964). Metabolism of vitamin D. 1. Preparation of radioactive vitamin D and its intestinal absorption in the rat. *J. clin. Invest.* **43**, 787–96.

Schaefer, J. W., Griffen, W. O. & Dubilier, L. D. (1968). Colonic lymphangiectasis associated with a potassium depletion syndrome. *Gastroenterol.* **55**, 515–21.

Schaffner, F. & Popper, H. (1963). Capillarization of hepatic sinusoids in man. *Gastroenterol.* **44**, 239–42.

Schiller, W. R., Suriyapa, C., Mutchler, J. H. W., Chen, S.-C. & Anderson, M. C. (1972). Controlled ductal infusion and absorption from the interstitium of canine pancreas. *Arch. Surg.* **105**, 356–62.

Schipp, R. (1967). Structure and ultrastructure of mesenteric lymphatic vessels. In *New Trends in Basic Lymphology*, ed. J. M. Collette, G. Jantet & E. Schoffeniels, pp. 50–7. Birkhauser, Basle & Stuttgart.

Schoefl, G. I. (1968). The ultrastructure of chylomicra and of the particles in an artificial fat emulsion. *Proc. R. Soc.* **169B**, 147–52.

References

Scholander, P. F., Hargens, A. R. & Miller, S. L. (1968). Negative pressure in the interstitial fluid of animals. *Science* **161**, 321–8.

Schønheyder, F. & Volqvartz, K. (1946). The gastric lipase in man. *Acta Physiol. Scand.* **11**, 349–60.

Schumaker, V. N. & Adams, G. H. (1969). Circulating lipoproteins. *Ann. Rev. Biochem.* **38**, 113–36.

Scow, R. O., Stein, Y. & Stein, O. (1967). Incorporation of dietary lecithin and lysolecithin into lymph chylomicrons in the rat. *J. biol. Chem.* **242**, 4919–24.

Seifert, J., Pröls, H., Messmer, K., Bücklein, R., Lob, G., Mehnert, H. & Brendel, W. (1975). Concentration and transport of different sugars in the lymph of the thoracic duct and in blood of the portal vein in dogs after enteral and parenteral administration. *Digestion* **12**, 221–31.

Semb, L. S., Rheault, M. J., Stevenson, J. K., Fletcher, T. L., Harkins, H. N. & Nyhus, L. M. (1964). A gastric inhibitory substance from dog antral juice. *Surg. Forum* **15**, 319–21.

Senior, J. R. & Isselbacher, K. J. (1963). Demonstration of an intestinal monoglyceride lipase; an enzyme with a possible role in the intracellular completion of fat digestion. *J. clin. Invest.* **42**, 187–95.

Serenyi, P. Magyar, Z. & Szabó, G. (1976). Cervical lymphatico-venous shunt in treatment of ascites in caval-constricted dogs and in patients with hepatic cirrhosis. *Lymphol.* **9**, 53–61.

Servelle, M., Bouvrain, Y., Tricot, R., Soulie, J., Turpyn, H., Frentz, F., Cornu, C. & Nadim, C. (1966). Lymphatic circulation in constrictive pericarditis. *J. cardiovasc. Surg.* **7**, 182–200.

Sessions, J. T., Viegas de Andrade, S. R. & Kokas, E. (1968). Intestinal villi: form and motility in relation to function. *Prog. Gastroenterol.* **1**, 248–60.

Shafiroff, B. G. P., Doubilet, H. & Ruggiero, W. (1939). Bilirubin resorption in obstructive jaundice. *Proc. Soc. exp. Biol. Med.* **42**, 203–5.

Shannon, A. D. & Lascelles, A. K. (1968a). The intestinal and hepatic contributions to the flow and composition of thoracic duct lymph in young milk-fed calves. *Q. J. exp. Physiol.* **53**, 194–205.

Shannon, A. D. & Lascelles, A. K. (1968b). Lymph flow and protein composition of thoracic duct lymph in the new-born calf. *Q. J. exp. Physiol.* **53**, 415–21.

Sheppard, M. S. & Sterns, E. E. (1974). The role of vascular and lymph capillaries in the clearance of interstitially injected albumin in acute and chronic inflammation of the colon. *Surg. Gynecol. Obstet.* **139**, 707–11.

Sheppard, M. S. & Sterns, E. E. (1975). The difference in clearance of interstitial albumin by the lymphatics from the stomach and the small and large intestine. *Surg. Gynec. Obstet.* **140**, 405–8.

Shim, W. K. T., Pollack, E. L. & Drapnas, J. (1961). Effect of serotonin, epinephrine, histamine and hexamethonium on thoracic duct lymph. *Am. J. Physiol.* **201**, 81–4.

Shrewsbury, M. M. (1959). Thoracic duct lymph in unanaesthetized mouse. Method of collection, rate of flow and cell content. *Proc. Soc. exp. Biol. Med.* **101**, 492–4.

Shrewsbury, M. M. & Reinhardt, W. O. (1952). Comparative metabolic effects of ingestion of water or 1 per cent sodium chloride solution in the rat with a thoracic duct lymph fistula. *Am. J. Physiol.* **168**, 366–74.

Shrivastava, B. K., Redgrave, T. G. & Simmonds, W. J. (1967). The source of endogenous lipid in the thoracic duct lymph of fasting rats. *Q. J. exp. Physiol.* **52**, 305–12.

Sieber, S. M., Cohn, V. H. & Wynn, W. T. (1974). The entry of foreign compounds into the thoracic duct lymph of the rat. *Xenobiotica* **4**, 265–84.

Silvester, C. F. (1911). On the presence of permanent communications between the lymphatic and the venous system at the level of the renal veins in adult South American monkeys. *Am. J. Anat.* **12**, 447–60.

301

References

Sim, D. N., Duprez, A. & Anderson, M. C. (1966). Modifications apportées au débit et à la composition de la lymphe du canal thoracique par la pancréatite expérimentale chez le chien. *Acta Gastroenterol. Belg.* **29**, 235–47.

Simmonds, W. J. (1954). The effect of fluid, electrolyte and food intake on thoracic duct lymph flow in unanaesthetised rats. *Aust. J. exp. Biol. med. Sci.* **32**, 285–300.

Simmonds, W. J. (1955). Some observations on the increase in thoracic duct lymph flow during intestinal absorption of fat in unanaesthetized rats. *Aust. J. exp. Biol. med. Sci.* **33**, 305–13.

Simmonds, W. J. (1957). The relationship between intestinal motility and the flow and rate of fat output in thoracic duct lymph in unanaesthetized rats. *Q. J. exp. Physiol.* **42**, 205–21.

Simpson-Morgan, M. W. & Smeaton, T. C. (1972). The transfer of antibodies by neonates and adults. *Adv. vet. Sci. comp. Med.* **16**, 355–86.

Singh, H. & Appert, H. A. (1969). Effect of pilocarpine and pancreatectomy on amylase and lipase levels of serum and hepatic lymph. *Proc. Soc. exp. Biol. Med.* **130**, 1122–5.

Singh, H., Pepin, J., Appert, H. E., Pairent, F. W. & Howard, J. M. (1969). Amylase and lipase secretion in the hepatic and intestinal lymph. I. Studies involving acute pancreatectomy. *Ann. Surg.* **169**, 233–9.

Sjövall, J. & Åkesson, I. (1955). Intestinal absorption of taurocholic acid in the rat. *Acta physiol. Scand.* **34**, 273–8.

Smallwood, R. A., Jones, E. A., Craigie, A., Raia, S. & Rosenoer, V. M. (1968). The delivery of newly synthesized albumin and fibrinogen to the plasma in dogs. *Clin. Sci.* **35**, 35–43.

Smeaton, T. C., Cole, G. J., Simpson-Morgan, M. W. & Morris, B. (1969). Techniques for the long term collection of lymph from the unanaesthetized foetal lamb *in utero*. *Aust. J. exp. Biol. Med. Sci.* **47**, 565–72.

Smith, J. B., McIntosh, G. H. & Morris, B. (1970). The traffic of cells through tissues: a study of peripheral lymph in sheep. *J. Anat.* **107**, 87–100.

Smith, R. O. (1949). Lymphatic contractility. A possible intrinsic mechanism of lymphatic vessels for the transport of lymph. *J. exp. Med.* **90**, 497–509.

Smith, T. (1925). Hydropic stages in the intestinal epithelium of newborn calves. *J. exp. Med.* **41**, 81–8.

Sotgiu, G., Cavalli, G. & Gasbarrini, G. (1963). Aspects morphologiques des petits vaisseaux et de l'épithélium du grêle dans l'hypertension portale expérimentale. *Arch. Malad. App. Dig. Nutr.* **52**, 739–50.

Stalder, H. & Joliat, G. (1973). Blood in the lymph (hemochylia) and intestinal lymphangiectasia associated with Lutembacher's syndrome. *Am. J. Med.* **55**, 99–104.

Starling, E. H. (1894). The influence of mechanical factors on lymph production. *J. Physiol. Lond.* **16**, 224–67.

Starling, E. H. (1896). On the absorption of fluids from the connective tissue spaces. *J. Physiol. Lond.* **19**, 312–26.

Starling, E. H. (1898). Production and absorption of lymph. In *Textbook of Physiology*, ed. E. A. Shäfer, vol. 1, pp. 285–311. Caxton, London.

Starling, E. H. (1909). *The Mercers' Company Lectures on the Fluids of the Body.* Constable, London.

Stassoff, B. (1914). Experimentelle Untersuchungen über die Kompensatorischen Vorgänge bei Darmresektion. *Beitr. Z. Klin. Chir.* **89**, 527–86.

Stein, Y. & Stein, O. (1966). Metabolism of labeled lysolecithin, lysophosphatidyl ethanolamine and lecithin in the rat. *Biochim. Biophys. Acta* **116**, 95–107.

Stephens, R. E. & Edds, K. T. (1976). Microtubules; structure, chemistry and function. *Phys. Rev.* **56**, 709–77.

Sternlieb, I., van den Hamer, C. J. A., & Alpert, S. (1967). Role of intestinal lymphatics in copper absorption. *Nature, Lond.* **216**, 824.

References

Sterns, E. E. & Vaughan, G. E. R. (1970). The lymphatics of the dog colon. *Cancer* **26**, 218–31.

Stoelinga, G. B., Van Munster, J. J. & Slooff, J. P. (1963). Chylous effusions into intestine in a patient with protein-losing gastroenteropathy. *Pediat.* **31**, 1011–18.

Strasberg, S. M., Dorn, B. C., Small, D. M. & Egdahl, R. H. (1971). The effect of biliary tract pressure on bile flow, bile salt secretion and bile salt synthesis in the primate. *Surg.* **70**, 140–6.

Straus, E., Gerson, C. D. & Yalow, R. S. (1974). Hypersecretion of gastrin associated with the short bowel syndrome. *Gastroenterol.* **66**, 175–80.

Strauss, E. W. (1966). Electron microscopic study of intestinal fat absorption *in vitro* from mixed micelles containing linolenic acid, monoolein and bile salt. *J. Lipid Res.* **7**, 307–23.

Strober, W., Wochner, R. D., Carbone, P. P. & Waldmann, T. A. (1967). Intestinal lymphangiectasia: a protein-losing enteropathy with hypogammaglobulinaemia, lymphocytopenia and impaired homograft rejection. *J. clin. Invest.* **46**, 1643–56.

Sudler, M. T. (1901). The architecture of the gall bladder. *Bull. Johns Hopkins Hosp.* **12**, 126–8.

Svanvik, J. (1973). Mucosal hemodynamics in the small intestine of the cat during regional sympathetic vasoconstrictor activation. *Acta physiol. Scand.* **89**, 19–29.

Svatŏs, A., Bartŏs, V. & Brzek, V. (1964). The concentration of cholecystokinin in human lymph and serum. *Arch. int. Pharmacodyn.* **149**, 515–20.

Sylven, C. (1970). Influence of blood supply on lipid uptake from micellar solutions by the rat small intestine. *Biochim. Biophys. Acta* **203**, 365–75.

Sylven, C. & Borgström, B. (1968). Absorption and lymphatic transport of cholesterol in the rat. *J. Lipid Res.* **9**, 596–601.

Szabó, G., Magyar, Z. & Jakab, F. (1975a). Bile constituents in blood and lymph during biliary obstruction. 1. The dynamics of absorption and transport of ions and colloid molecules. *Lymphol.* **8**, 29–36.

Szabó, G., Magyar, Z. & Jakab, F. (1975b). The lymphatic drainage of the liver capsula and hepatic parenchyma. *Res. exp. Med.* **166**, 193–200.

Szabó, G., Magyar, Z., Szentirmai, A., Jakab, F. & Mihaly, K. (1975). Bile constituents in blood and lymph during biliary obstruction. II. The absorption and transport of bile acids and bilirubin. *Lymphol.* **8**, 36–42.

Takashima, T. & Takekoshi, N. (1968). Lymphographic evaluation of abnormal lymph flow in protein-losing gastroenteropathy secondary to chronic constrictive pericarditis. *Radiol.* **90**, 502–6.

Tasker, R. R. (1951). The collection of intestinal lymph from normally active rats. *J. Physiol., Lond.* **115**, 292–5.

Taylor, A. E., Gibson, H. & Gaar, K. A. Jr. (1970). Effect of tissue pressure on lymph flow. *Biophys. J.* **10**, 45A.

Taylor, A. E., Gibson, W. H., Granger, H. J. & Guyton, A. C. (1973). The interaction between intracapillary and tissue forces in the overall regulation of interstitial fluid volume. *Lymphol.* **6**, 192–208.

Taylor, K. B. & French, J. E. (1960). The role of the lymphatics in the intestinal absorption of vitamin B_{12} in the rat. *Q. J. exp. Physiol.* **45**, 72–6.

Thompson, G. R., Ockner, R. K. & Isselbacher, K. J. (1969). Effect of mixed micellar lipid on the absorption of cholesterol and vitamin D_3 into lymph. *J. clin. Invest.* **48**, 87–95.

Thompson, S. Y., Braude, R., Coates, M. E., Cowie, A. T., Ganguly, J. & Kon, S. K. (1950). Further studies of the conversion of β-carotene to vitamin A in the intestine. *Br. J. Nutr.* **4**, 398–421.

Threefoot, S. A. (1968). Gross and microscopic anatomy of the lymphatic vessels and lymphatico-venous communications. *Cancer Chemotherap. Rep.* **52**, 1–20.

303

References

Threefoot, S. A. & Kossover, M. F. (1966). Lymphaticovenous communications in man. *Arch. Int. Med.* **117**, 213–23.

Threefoot, S. A., Kent, W. T. & Hatchett, B. F. (1963). Lymphatico-venous and lymphatico-lymphatic communications demonstrated by plastic corrosion models of rats and by postmortem lymphangiography in man. *J. lab. clin. Med.* **61**, 9–22.

Tift, W. L. & Lloyd, J. K. (1975). Intestinal lymphangiectasia. Long term results with MCT diet. *Arch. Dis. Child.* **50**, 269–76.

Tilney, N. L. (1971). Patterns of lymphatic drainage in the adult laboratory rat. *J. Anat.* **109**, 369–83.

Tilney, N. L. & Murray, J. E. (1968). Chronic thoracic duct fistula: operative technic and physiologic effects in man. *Ann. Surg.* **167**, 1–8.

Tirone, P., Schiantarelli, P. & Rosati, G. (1973). Pharmacological activity of some neuro-transmitters in the isolated thoracic duct of dogs. *Lymphol.* **6**, 65–8.

Traber, D. L., Haynes, J., Daily, L. J. & Kolmen, S. N. (1963). Coagulation defects as a result of chronic lymphatic diversion. *Texas Rep. Biol. Med.* **21**, 587–600.

Treadwell, C. R. & Vahouny, G. V. (1968). Cholesterol absorption. In *Handbook of Physiology*, Section 6, *Alimentary Canal*, Vol. III. *Intestinal Absorption*. American Physiological Society, Washington, 1407–38.

Turner, S. G. & Barrowman, J. A. (1977). Intestinal lymph flow and lymphatic transport of protein during fat absorption. *Q. J. exp. Physiol.* **62**, 175–80.

Ulstrom, R. A. & Krivit, W. (1960). Exudative enteropathy, hypoproteinaemia, edema and iron deficiency anaemia. *Am. J. Dis. Child.* **100**, 509–12.

Vaerman, J.-P. André, C., Bazin, H. & Heremans, J. F. (1973). Mesenteric lymph as a major source of serum IgA in guinea pigs and rats. *Europ. J. Immunol.* **3**, 580–4.

Vaerman, J.-P. & Heremans, J. F. (1970). Origin and molecular size of immunoglobulin-A in the mesenteric lymph of the dog. *Immunol.* **18**, 27–38.

Vahouny, G. V. & Treadwell, C. R. (1964). Absolute requirement for free sterol for absorption by rat intestinal mucosa. *Proc. Soc. exp. Biol. Med.* **116**, 496–8.

Vajda, J. (1966). Innervation of lymph vessels. *Acta Morph. Acad. Sci. Hung.* **14**, 197–208.

Vajda, J. & Leranth, C. (1967). The lymphatic system of the large intestine. *Acta Morph. Acad. Sci. Hung.* **15**, 257–63.

Vajda, J. & Tömböl, T. (1965). Lymphatic apparatus of the wall of the small intestine. *Acta Morph. Acad. Sci. Hung.* **13**, 339–47.

Vanlerenberghe, J., Trupin, N., Nguyen, M. & Rose, A. (1972). Absorption intestinale de la brome sulfone phtaléine. Rôle de la voie lymphatique. *Compt. Rend. Soc. Biol.* **166**, 613–18.

Vardy, P., Lebenthal, E. & Schwachman, H. (1975). Intestinal lymphangiectasia. A reappraisal. *Pediatr.* **55**, 842–51.

Vatner, S. F., Franklin, D. & Van Citters, R. L. (1970). Mesenteric vasoactivity associated with eating and digestion in the conscious dog. *Am. J. Physiol.* **219**, 170–4.

Vega, R. E., Appert, H. E. & Howard, J. M. (1967). The effects of secretin in stimulating the output of amylase and lipase in the thoracic duct of the dog. *Ann. Surg.* **166**, 995–1001.

Vescia, F. G. & Davis, J. H. (1965). Treatment of intestinal lymphangiectasia. *Gastroenterol.* **48**, 287.

Viegas de Andrade, S. R., Bozymski, E. M. & Sessions, J. T. (1968). A method for cannulating mesenteric lymphatic vessels in dogs. *J. Lab. clin. Med.* **72**, 521–6.

Volkheimer, G., Schulz, F. H., Lindenau, A. & Beitz, U. (1969). Persorption of metallic iron particles. *Gut* **10**, 32–3.

Volwiler, W., Grindlay, J. H. & Bollman, J. L. (1950). Symposium on liver disease; relation of portal vein pressure to the formation of ascites – an experimental study. *Gastroenterol.* **14**, 40–55.

Waldmann, T. A. (1966). Protein-losing enteropathy. *Gastroenterol.* **50**, 422–43.

Waldmann, T. A. & Schwab, P. J. (1965). IgG (7 S-gammaglobulin) metabolism in hypogammaglobulinemia. Studies in patients with defective gammaglobulin synthesis, gastrointestinal protein loss, or both. *J. clin. Invest.* **44**, 1523–33.

Waldmann, T. A., Steinfeld, J. L., Dutcher, T. F., Davidson, J. D. & Gordon, R. S. (1961). The role of the gastrointestinal system in 'idiopathic' hypoproteinemia. *Gastroenterol.* **41**, 197–207.

Wallentin, I. (1966). Importance of tissue pressure for the fluid equilibrium between the vascular and interstitial compartments in the small intestine. *Acta physiol. Scand.* **68**, 304–15.

Wang, C. A., Caro, A. & Yamazaki, Z. (1969). A technique for the prolonged intermittent sampling of hepatic lymph. An exteriorized catheter lymphatic venous shunt in the dog. *Surg.* **65**, 783–8.

Warnock, M. L. (1968). Intestinal phosphatase and fat absorption. *Proc. Soc. exp. Biol. Med.* **129**, 768–72.

Warren, W. D., Fomon, J. J. & Leite, C. A. (1968). Critical assessment of the rationale of thoracic duct drainage in the treatment of portal hypertension. *Surg.* **63**, 7–16.

Warshaw, A. L. & Walker, W. A. (1974). Intestinal absorption of intake antigenic protein. *Surg.* **76**, 495–9.

Warshaw, A. L., Walker, W. A., Cornell, R. & Isselbacher, K. J. (1971). Small intestinal permeability to macromolecules. Transmission of horse radish peroxidase into mesenteric lymph and portal blood. *Lab. Inv.* **25**, 675–84.

Warshaw, A. L., Walker, W. A. & Isselbacher, K. J. (1974). Protein uptake by the intestine: evidence for absorption of intact macromolecules. *Gastroenterol.* **66**, 987–92.

Watanabe, K. & Fishman, W. H. (1964). Application of the stereospecific inhibitor L-phenylalanine to the enzymorphology of intestinal alkaline phosphatase. *J. Histochem. Cytochem.* **12**, 252–60.

Watkins, A. L. & Fulton, M. N. (1938). The effect of fluids given intraperitoneally, intravenously and by mouth on volume of thoracic duct lymph in dogs. *Am. J. Physiol.* **122**, 281–7.

Webb, R. L. & Nicoll, P. A. (1944). Behaviour of lymphatic vessels in the living bat. *Anat. Rec.* **88**, 351–67.

Webb, R. L. & Nicoll, P. A. (1951). Persistence of active vasomotion along blood and lymphatic vessels in the bat's wing after denervation. *Anat. Rec.* **109**, 414.

Weiner, S., Gramatica, L., Voegle, L. D., Hauman, R. L. & Anderson, M. C. (1970). Role of the lymphatic system in the pathogenesis of inflammatory disease in the biliary tract and pancreas. *Am. J. Surg.* **119**, 55–61.

Weinstein, L. D., Scanlon, G. T. & Hersh, J. (1969). Chylous ascites: Management with medium-chain triglycerides and exacerbation by lymphangiography. *Amer. J. dig. Dis.* **14**, 500–9.

Weiss, J. M. (1955). The role of the Golgi complex in fat absorption as studied with the electron microscope with observations on the cytology of the duodenal absorptive cells. *J. exp. Med.* **102**, 775–82.

Wells, H. S. (1932). The concentration and osmotic pressure of the proteins in blood serum and in lymph from the lacteals of dogs. *Am. J. Physiol.* **101**, 421–33.

Werner, B. (1965). Thoracic duct cannulation in man. I. Surgical technique and a clinical study on 79 patients. *Acta chir. Scand.* **Suppl. 353**.

Werner, B. (1966a). The biochemical composition of the human thoracic duct lymph. *Acta chir. Scand.* **132**, 63–76.

Werner, B. (1966b). Thoracic duct cannulation in man. II. A follow-up study on 22 patients. *Acta chir. Scand.* **132**, 93–105.

References

Whipple, G. H. & King, J. H. (1911). The pathogenesis of icterus. *J. exp. Med.* **13**, 115–35.

Whyte, M., Goodman, D. S. & Karmen, A. (1965). Fatty acid esterification and chylomicron formation during fat absorption in the rat. 3. Positional relations in triglycerides and lecithin. *J. Lipid Res.* **6**, 233–40.

Wiederhielm, C. A. (1967). Analysis of small vessel function. In *Physical Bases of Circulatory Transport: Regulation and Exchange*, ed. E. B. Reeve & A. C. Guyton, p. 313. W. B. Saunders Co., Philadelphia.

Wilson, F. A., Sallee, V. L. & Dietschy, J. M. (1971). Unstirred water layers in intestine: rate determinant of fatty acid absorption from micellar solutions. *Science* **174**, 1031–3.

Wilson, T. H. (1956). A modified method for study of intestinal absorption in vitro. *J. appl. Physiol.* **9**, 137–40.

Windmueller, H. G., Herbert, P. N. & Levy, R. I. (1973). Biosynthesis of lymph and plasma lipoprotein apoproteins by isolated perfused rat liver and intestine. *J. Lipid Res.* **14**, 215–23.

Winkenwerder, W. L. (1927). A study of the lymphatics of the gall bladder of the cat. *Bull. Johns Hopkins Hosp.* **41**, 226–38.

Wisse, E. (1970). An electron microscopic study of the fenestrated endothelial lining of rat liver sinusoids. *J. ultrastruct. Res.* **31**, 125–50.

Witte, C. L., Cole, W. R., Clauss, R. H. & Dumont, A. E. (1968). Splanchnic tissue oxygenation: estimation by thoracic duct lymph pO_2. *Lymphol.* **1**, 109–16.

Witte, C. L., Witte, M. H., Cole, W. R., Chung, Y. C., Bleisch, V. R. & Dumont, A. E. (1969). Dual origin of ascites in hepatic cirrhosis. *Surg. Gynec. Obstet.* **129**, 1027–33.

Witte, C. L., Witte, M. H., Bair, G., Mobley, W. P. & Morton, D. (1974). Experimental study of hyperdynamic vs stagnant mesenteric blood flow in portal hypertension. *Ann. Surg.* **179**, 304–10.

Witte, C. L., Witte, M. H., Dumont, A. E., Frist, J. & Cole, W. R. (1968). Lymph protein in hepatic cirrhosis and experimental hepatic and portal venous hypertension. *Ann. Surg.* **168**, 567–77.

Witte, M. H., Dumont, A. E., Levine, N. & Cole, W. R. (1968). Patterns of distribution of sulfobromophthalein in lymph and blood during obstruction to bile flow. *Am. J. Surg.* **115**, 69–74.

Witte, M. H., Dumont, A. E., Clauss, R. H., Rader, B., Levine, N. & Breed, E. S. (1969a). Lymph circulation in congestive heart failure: Effect of external thoracic duct drainage. *Circ.* **39**, 723–33.

Witte, M. H., Dumont, A. E., Cole, W. R., Witte, C. L. & Kintner, K. (1969b). Lymph circulation in hepatic cirrhosis: effect of portacaval shunt. *Ann. intern. Med.* **70**, 303–10.

Wollin, A. & Jaques, L. B. (1973). Plasma protein escape from the intestinal circulation to the lymphatics during fat absorption. *Proc. Soc. exp. Biol. Med.* **142**, 1114–17.

Wollin, A. & Jaques, L. B. (1974). Increased diamine oxidase activity in the intestinal lymph of rats on fat ingestion. *Can. J. Physiol. Pharmacol.* **52**, 760–2.

Wollin, A. & Jaques, L. B. (1976). Blocking of olive oil induced plasma protein escape from the intestinal circulation by histamine antagonists and by a diamine oxidase releasing agent. *Agents and Actions* **6**, 589–92.

Woolley, G. & Courtice, F. C. (1962). The origin of albumin in hepatic lymph. *Austr. J. exp. Biol. Med. Sci.* **40**, 121–8.

Woytkiw, L. & Esselbaugh, N. C. (1951). Vitamin A and carotene absorption in the guinea pig. *J. Nutr.* **43**, 451–8.

Yablonski, M. E. & Lifson, N. (1976). Mechanism of production of intestinal secretion by elevated venous pressure. *J. clin. Invest.* **57**, 904–15.

References

Yakimets, W. W. & Bondar, G. F. (1966). Thoracic duct lymph in the control of gastric secretion. *Surg. Forum* **17**, 307–9.

Yakimets, W. W. & Bondar, G. F. (1967). The effect of complete thoracic duct lymph diversion on gastric secretion in dogs. *Canad. J. Surg.* **10**, 218–22.

Yakimets, W. W. & Bondar, G. F. (1969). Hormonal stimulatory mechanism producing gastric hypersecretion following massive small intestinal resection. *Canad. J. Surg.* **12**, 241–4.

Yalow, R. S. (1975). Heterogeneity of peptide hormones with relation to gastrin. In *Gastro-intestinal Hormones*, ed. J. C. Thompson. pp. 25–41. University of Texas Press, Austin.

Yoffey, J. M. & Courtice, F. C. (1956). *Lymphatics, Lymph and Lymphoid Tissue*. Arnold, London.

Yoffey, J. M. & Courtice, F. C. (1970). *Lymphatics, Lymph and the Lymphomyeloid Complex*. Academic Press, New York & London.

Yokoyama, A. & Zilversmit, D. B. (1965). Particle size and composition of dog lymph chylomicrons. *J. Lipid Res.* **6**, 241–6.

Zemel, R. & Gutelius, J. R. (1965). Anatomy and function of the thoracic duct-venous junction. *Surg. Forum* **16**, 138–9.

Zeppa, R. & Womack, N. A. (1963). Humoral control of hepatic lymph flow. *Surg.* **54**, 37–44.

Zhdanov, D. A. (1953). Anastomoses and confluences of the lymphatic system of stomach and duodenum of man. Cited by Rusznyak, I., Földi, M. & Szabó, G. (1967). In *Lymphatics and Lymph Circulation. Physiology and Pathology*, 2nd edn. Pergamon Press, Oxford.

Zilversmit, D. B. (1965). The composition and structure of lymph chylomicrons in dog, rat and man. *J. clin. Invest.* **44**, 1610–22.

Zilversmit, D. B. (1968). The surface coat of chylomicrons: lipid chemistry. *J. Lipid Res.* **9**, 180–6.

Zilversmit, D. B., Sisco, P. H. & Yokoyama, A. (1966). Size distribution of thoracic duct lymph chylomicrons from rats fed cream and corn oil. *Biochim. Biophys. Acta* **125**, 129–35.

Zimmon, D. S., Oratz, M., Kessler, R., Schreiber, S. S. & Rothschild, M. A. (1969). Albumin to ascites: demonstration of a direct pathway bypassing the systemic circulation. *J. clin. Invest.* **48**, 2074–8.

Zweifach, B. W. & Prather, J. W. (1975). Micromanipulation of pressure in terminal lymphatics in the mesentery. *Am. J. Physiol.* **228**, 1326–35.

Supplementary references

Baxter, J. H., Steinberg, D., Mize, C. E. & Avigan, J. (1967). Absorption and metabolism of uniformly ^{14}C-labelled phytol and phytanic acid by the intestine of the rat studied with thoracic duct cannulation. *Biochim. Biophys. Acta* **137**, 277–90.

Biber, B., Fara, J. & Lundgren, O. (1973b). Intestinal vascular responses to 5-hydroxytryptamine. *Acta Physiol. Scand.* **87**, 526–34.

Borgström, B. (1974). Fat digestion and absorption. *Biomembranes*, **4B**, 555–620.

Peake, I. R., Windmueller, H. G. & Bieri, J. G. (1972). A comparison of the intestinal absorption, lymph and plasma transport and tissue uptake of α- and γ-tocopherols in the rat. *Biochim. Biophys. Acta* **260**, 679–88.

Pocock, D. M. E. & Vost, A. (1974). DDT absorption and chylomicron transport in the rat. *Lipids*, **9**, 374–81.

Razin, E., Feldman, M. G. & Dreiling, D. A. (1962). The hormonal regulation of thoracic ductal lymph flow. *J. Surg. Res.* **2**, 320–31.

Savary, P. & Constantin, M. J. (1967). Sur la solubilisation micellaire de l'hexadecane et son passage dans la lymphe thoracique du rat. *Biochim. Biophys. Acta* **137**, 264–76.

INDEX

Index

Index

prelymphatics, 10
prostaglandins in lymph, 79
protein absorption, intestinal, lymphatic transport, 146–55
protein-losing gastroenteropathy
  in constrictive pericarditis, 265–8
  in intestinal lymphangiectasia, 261
  techniques for demonstration, 260

schistosomiasis, hepatic fibrosis and lymph flow, 245–7
secretin, effect on intestinal blood flow, 49–51
β-sitosterol, 200, 202
space of Disse, 229–31
sphingomyelins, absorption, 211
Starling principle, 31–5
steroid hormones, absorption and lymphatic transport, 162

thoracic duct, 7
  cannulation, 85
  constrictive pericarditis, 268
  drainage in portal hypertension and ascites, 247–51
  ligation, 14, 92, 270
  schistosomal hepatic fibrosis, 246
  –venous shunts, 86
thoracic duct–jugular venous junction, 6, 10, 248–51

thoracic pump, lymph flow, 74
tissue injury, lymph flow and composition, 79
tocopherols, see vitamin E
toxins, transport in lymph, 62, 146
transaminases, in lymph, 61, 79, 106
tributyrinase, see lipase
triglycerides, digestion and absorption, 167–71
trypsin, in acute pancreatitis, 113

ubiquinones, absorption and lymphatic transport, 227

VLDL, see lipoproteins
vasopressin, effect on lymph flow, 69
villus pump, 25, 74
vitamin A
  intestinal absorption, 213–17
  portal: lymphatic partition, 199, 216
vitamin B_{12}, lymphatic transport, 158
vitamin D, intestinal absorption, 217–20
vitamin E, intestinal absorption, 220–4
vitamin K
  transport in lymph, 88
  intestinal absorption, 224–6

water absorption, lymphatic transport, 128–42
Whipple's disease, 268–70

312